Mathematical Engineering

Series editors

Today, the development of high-tech systems is unthinkable without mathematical modeling and analysis of system behavior. As such, many fields in the modern engineering sciences (e.g. control engineering, communications engineering, mechanical engineering, and robotics) call for sophisticated mathematical methods in order to solve the tasks at hand.

The series Mathematical Engineering presents new or heretofore little-known methods to support engineers in finding suitable answers to their questions, presenting those methods in such manner as to make them ideally comprehensible and applicable in practice.

Therefore, the primary focus is—without neglecting mathematical accuracy—on comprehensibility and real-world applicability.

To submit a proposal or request further information, please use the PDF Proposal Form or contact directly: *Dr. Jan-Philip Schmidt, Publishing Editor (jan-philip.schmidt@springer.com).*

More information about this series at http://www.springer.com/series/8445

Jonathan Whiteley

Finite Element Methods

A Practical Guide

Springer

Jonathan Whiteley
Department of Computer Science
University of Oxford
Oxford
UK

Additional material to this book can be downloaded from http://extras.springer.com.

ISSN 2192-4732 ISSN 2192-4740 (electronic)
Mathematical Engineering
ISBN 978-3-319-49970-3 ISBN 978-3-319-49971-0 (eBook)
DOI 10.1007/978-3-319-49971-0

Library of Congress Control Number: 2016958482

Printed on acid-free paper

This Springer imprint is published by Springer Nature
The registered company is Springer International Publishing AG
The registered company address is: Gewerbestrasse 11, 6330 Cham, Switzerland

Preface

This book has evolved from courses on the finite element method that have been taught as part of several graduate programmes at the University of Oxford. These courses have all focused on the practical application of the finite element method. A typical course would begin with a specific class of differential equations being written down. The course would then provide students with all the detail that is required to develop a computational implementation of the finite element method for this class of differential equations. This approach requires the following topics to be addressed: (i) derivation of the weak formulation of a differential equation; (ii) discretisation of the domain on which the differential equation is defined into elements; (iii) specification of suitable basis functions; (iv) derivation of a system of algebraic equations that is derived from the finite element formulation of the differential equation; (v) solution of this system of algebraic equations; and (perhaps most importantly) (vi) the practical implementation of all steps. This book is written in the spirit in which these courses have been taught. Mathematical rigour is certainly needed when discussing some of these topics, such as the use of Sobolev spaces when deriving the weak formulation of a differential equation. The focus, however, is on the practical implementation of the finite element method for a given differential equation.

This book begins with a brief overview of the finite element method in Chap. 1. Chapters 2–6 then focus on the application of the finite element method to ordinary differential equations. In Chap. 2, we begin with a very simple example differential equation. Although simple, this example allows illustration of the key ideas that underpin the finite element method. We then consider general, linear ordinary differential equations in Chap. 3, providing a more rigorous discussion than in Chap. 2. The material in Chap. 3 focuses exclusively on finite element solutions that are a linear approximation to the true solution on each element. These concepts are extended in Chap. 4 to explain how a higher order polynomial approximation may be used on each element. We then explain how nonlinear ordinary differential equations are handled by the finite element method in Chap. 5, and how systems of ordinary differential equations are handled in Chap. 6. In all of these chapters, we carefully explain how the values of the solution at each node of the computational mesh satisfy a system of algebraic equations. We explain how these algebraic equations may be assembled, and how they may be solved. Each chapter contains an exemplar computational implementation in the form of MATLAB functions.

Chapters 7–12 focus on the application of the finite element method to elliptic partial differential equations, and these chapters take a very similar approach to that used in earlier chapters. We begin with a simple example in Chap. 7, before considering more general linear elliptic equations in Chap. 8. In both of these chapters, we partition the domain on which the differential equation is defined into triangular elements. In Chap. 9, we explain how the material presented in Chap. 8 may be modified to allow the use of a mesh that arises from partitioning the domain into quadrilateral elements. Higher order polynomial approximations to the solution are considered in Chap. 10, followed by nonlinear partial differential equations in Chap. 11 and systems of elliptic equations in Chap. 12. Exemplar computational implementations of the finite element method are given in several of these chapters. Finally, in Chap. 13, we explain how the finite element method may be applied to parabolic partial differential equations.

Every application of the finite element method in this book requires the solution of a system of algebraic equations. Appendix A provides a brief summary of computational techniques that are available for solving both linear and nonlinear systems of algebraic equations. This summary covers both direct and iterative methods for solving linear systems, preconditioners that accelerate the convergence of iterative methods for solving linear systems, and iterative methods for solving nonlinear systems.

My understanding and appreciation of the finite element method has been substantially enhanced by discussions with colleagues too numerous to mention. Special mention must, however, be made of both the Computational Biology Group in the Department of Computer Science at the University of Oxford, and the Numerical Analysis Group in the Mathematical Institute at the University of Oxford. The finite element research interests of these two groups cover a wide spectrum, addressing a range of both practical and theoretical issues. I am very fortunate to have been a member of both of these research groups for periods of time over the last two decades.

Above all, I would like to thank my family for their support during the writing of this book.

Oxford, UK
April 2016

Jonathan Whiteley

Contents

Chapter 1
An Overview of the Finite Element Method

Differential equations play a central role in quantitative sciences and are used to describe phenomena as diverse as the flow of traffic through a city, the evolution of an oilfield over many years, the flow of ions through an electrochemical solution and the effect of a drug on a given patient. Most differential equations that are encountered in practice do not possess an analytic solution, that is a solution that can be written in terms of known functions. Under these circumstances, a numerical solution to the differential equation is often sought. The finite element method is one common technique for calculating a numerical solution of a differential equation.

The underlying principle of the finite element method is very simple to express. The domain on which the differential equation is defined is partitioned into many smaller regions known as *elements*, and the solution of the differential equation is approximated on each of these elements using a low-order polynomial function. Consider the example partial differential equation, defined on the interior of the ellipse $x^2/4 + y^2 = 1$ by:

$$-\left(\frac{\partial^2 u}{\partial x^2} + \frac{\partial^2 u}{\partial y^2}\right) = 1.$$

We apply Dirichlet boundary conditions on the boundary of the ellipse where $y \leq 0$, given by

$$u(x, y) = x,$$

and Neumann boundary conditions on the boundary of the ellipse where $y > 0$, given by

$$(\nabla u) \cdot \mathbf{n} = \sin(2(x + y)).$$

J. Whiteley, *Finite Element Methods*, Mathematical Engineering,
DOI 10.1007/978-3-319-49971-0_1

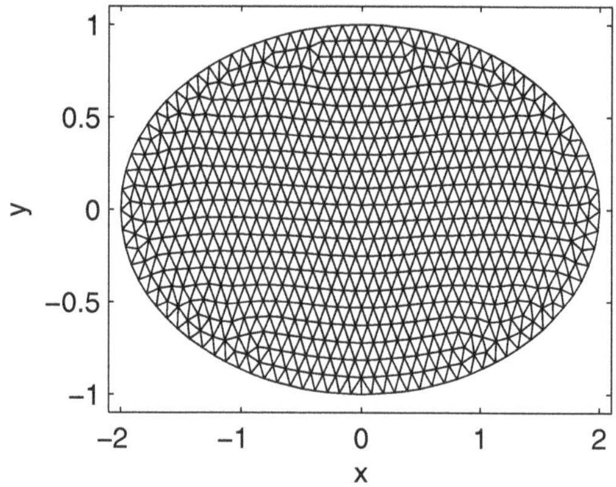

Fig. 1.1 The partitioning of the domain $x^2/4 + y^2 < 1$ into *triangular* elements

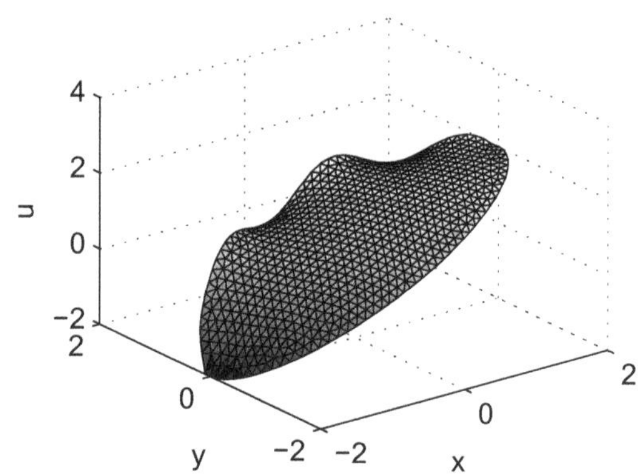

Fig. 1.2 The finite element approximation to the solution of the differential equation

The vector **n** that appears in the specification of the Neumann boundary conditions is the normal vector pointing out of the ellipse, with unit length.

To calculate the finite element solution of the differential equation above, we first partition the domain on which the differential equation is defined, namely the interior of the ellipse $x^2/4 + y^2 = 1$, into elements. We illustrate how this domain may be partitioned into triangular elements in Fig. 1.1.

Having partitioned the ellipse into triangular elements, we then calculate the finite element solution of the differential equation. The finite element solution that is a linear approximation to the solution on each element is shown in Fig. 1.2.

1.1 Features of the Finite Element Method

We will see in later chapters that the finite element method is a very flexible method for calculating the numerical solution of differential equations. In particular, it has three key features that make it very attractive from a practical viewpoint. First, the finite element method can handle domains, such as that shown in Fig. 1.1, that are of arbitrary shape. This is clearly a useful property, as in practice differential equations are posed on domains such as the exterior of an aircraft (when estimating the forces acting on the aircraft); a whole country (when modelling the spread of a disease); or a subject's lungs (when modelling oxygen transport from air into the subject's blood). The second key feature of the finite element method is that boundary conditions that include the derivative of the solution, such as the Neumann boundary conditions for the example differential equation presented earlier in this chapter, may be handled very easily. Finally, when we desire more accuracy in our finite element solution, we have two options—we may partition the domain into smaller triangular elements or increase the degree of the polynomial approximation on each element.

The finite element method also has attractive features when viewed from a theoretical standpoint. As we are approximating the true solution by a polynomial function on each element, functional analysis may be used to assess how well the finite element solution approximates the true solution. Very powerful theorems exist for estimating the error between the finite element solution and the true solution of a differential equation. These theorems have spawned practical techniques for guiding the partitioning of the domain so that the finite element solution is within a user supplied tolerance of the true solution.

1.2 Using This Book

This book is organised so that the application of the finite element method to ordinary differential equations is covered in Chaps. 2–6, and the application to partial differential equations is covered in Chaps. 7–13. For both ordinary differential equations and

Table 1.1 The prerequisite chapter(s) for each chapter in this book

Chapter	Prerequisite chapter(s)
2	1
3	1, 2
4	1, 2, 3
5	1, 2, 3
6	1, 2, 3
7	1, 2, 3
8	1, 2, 3,7
9	1, 2, 3, 7, 8
10	1, 2, 3, 4, 7, 8
11	1, 2, 3, 5, 7, 8
12	1, 2, 3, 6, 7, 8
13	1, 2, 3, 7, 8

partial differential equations, we first explain how to calculate finite element solutions that use a linear approximation to the solution on each element. Subsequent chapters then address the use of higher order approximations to the solution on each element; nonlinear differential equations; and systems of differential equations. Some readers may not want to read all chapters as they may, for example, only be interested in learning how to apply the finite element method to linear differential equations. In Table 1.1, we list the prerequisite chapters for each chapter of this book.

Many chapters include exemplar computational implementations that are written for use with MATLAB. The MATLAB functions are intended to aid the reader's understanding through being written in a way that is as close as possible to the algorithms that are derived earlier in the chapter. As a consequence, some fragments of these listings are written using more lines of code than is strictly necessary; some `for` loops, for example, could be written in a more elegant style using the vectorisation feature offered by MATLAB. The justification for this more verbose programming style is that it will allow readers who choose to use a programming language other than MATLAB to easily translate the listings provided into their programming language of choice. These MATLAB listings may be downloaded from http://extras.springer.com.

Chapter 2
A First Example

In Chap. 1, we explained the fundamental philosophy that underpins the finite element method—that is, the region on which a differential equation is defined is partitioned into smaller regions known as elements, and the solution on each of these elements is approximated using a low-order polynomial function. In this chapter, with the aid of a simple example, we illustrate how this may be done. This overview will require the definition of some terms that the reader may not be familiar with, as well as a few technical details. We will, however, undertake to keep these definitions to a minimum and will focus on the underlying strategy of applying the finite element method without getting bogged down by these technical details. As a consequence, we will inevitably skate over some mathematical rigour, but will make a note to return to these points in later chapters.

2.1 Some Brief Mathematical Preliminaries

We make a few comments on notation before we begin our overview of the finite element method. Throughout this book, we use the convention that a solution of a differential equation that is represented by a lowercase letter, for example $u(x)$, is the exact, or analytic, solution of the differential equation. The corresponding uppercase letter, $U(x)$ in this case, will represent the finite element approximation to the exact solution.

We will frequently make use of vectors and matrices, and adopt the following conventions.

1. Vectors will be assumed to be column vectors, i.e. a vector with many rows but only one column, and will be denoted by bold font. Entry i of the vector $\mathbf{x}$ will be denoted by x_i.

J. Whiteley, *Finite Element Methods*, Mathematical Engineering,
DOI 10.1007/978-3-319-49971-0_2

2. Matrices will be represented by upper case letters typeset in italic font. The entry in row i and column j of the matrix A will be denoted by $A_{i,j}$.
3. The indexing of both vectors and matrices will start from 1. A vector $\mathbf{b}$ with N entries, for example, will have entries $b_1, b_2, \ldots, b_N$. This allows the finite element algorithms to be written in a way that facilitates the writing of software in programming languages or environments such as MATLAB, Octave or Fortran, where array indexing starts from 1 (known as "one-based indexing"). Readers using programming languages such as C, C++ and Python, where the indexing of arrays starts from 0 (known as "zero-based indexing") will need to adapt the algorithms to take account of this feature of the programming language being used.

2.2 A Model Differential Equation

We will use a simple boundary value problem to demonstrate the application of the finite element method. A boundary value problem comprises both a differential equation and boundary conditions. In this chapter, we will consider only *Dirichlet boundary conditions*, where the value of the unknown function $u(x)$ is specified on the boundary. More complex boundary conditions, such as those that include the derivative of $u(x)$ on the boundary, will be discussed in Chap. 3. Our model boundary problem comprises the differential equation

$$-\frac{\mathrm{d}^2u}{\mathrm{d}x^2} = 2, \qquad 0 < x < 1, \tag{2.1}$$

and the Dirichlet boundary conditions given by

$$u(0) = u(1) = 0. \tag{2.2}$$

The region $0 < x < 1$ on which the equation is defined is known as the *domain*. This simple example, with solution $u(x) = x(1-x)$, will now be used to exhibit the underlying principles of calculating the finite element solution of a differential equation.

2.3 The Weak Formulation

Let $u(x)$ be the solution of the model differential equation, Eq. (2.1), subject to the boundary conditions given by Eq. (2.2). The function $u(x)$ is known as the *classical solution* of this differential equation, and specification of $u(x)$ through Eqs. (2.1) and (2.2) is known as the *classical formulation*.

We now introduce the *weak formulation* of the model problem, a concept that may be unfamiliar to some readers. In this chapter, we limit ourselves to illustrating the weak formulation of a differential equation by example, rather than giving a watertight definition through precise mathematical statements. A more complete and rigorous treatment of this material will be presented in Sect. 3.2.

To derive the weak formulation of our model problem, we first define $v(x)$ to be a continuous function that satisfies

$$v(0) = v(1) = 0, \tag{2.3}$$

so that $v(x) = 0$ at the values of x where we apply Dirichlet boundary conditions. We now multiply the differential equation, given by Eq. (2.1), by $v(x)$, and integrate the resulting product over the interval $0 < x < 1$ on which the differential equation is defined, giving

$$\int_0^1 -\frac{\mathrm{d}^2 u}{\mathrm{d}x^2} v(x)\,\mathrm{d}x = \int_0^1 2v(x)\,\mathrm{d}x.$$

We now apply integration by parts to the left-hand side of the equation above to obtain

$$\left[-\frac{\mathrm{d}u}{\mathrm{d}x} v(x)\right]_0^1 + \int_0^1 \frac{\mathrm{d}u}{\mathrm{d}x}\frac{\mathrm{d}v}{\mathrm{d}x}\,\mathrm{d}x = \int_0^1 2v(x)\,\mathrm{d}x,$$

which may be written as

$$-\left.\frac{\mathrm{d}u}{\mathrm{d}x}\right|_{x=1} v(1) + \left.\frac{\mathrm{d}u}{\mathrm{d}x}\right|_{x=0} v(0) + \int_0^1 \frac{\mathrm{d}u}{\mathrm{d}x}\frac{\mathrm{d}v}{\mathrm{d}x}\,\mathrm{d}x = \int_0^1 2v(x)\,\mathrm{d}x, \tag{2.4}$$

where

$$\left.\frac{\mathrm{d}u}{\mathrm{d}x}\right|_{x=1} \quad \text{and} \quad \left.\frac{\mathrm{d}u}{\mathrm{d}x}\right|_{x=0}$$

denote the value taken by $\frac{\mathrm{d}u}{\mathrm{d}x}$ at $x = 1$ and $x = 0$, respectively. Remembering our condition on $v(x)$, given by Eq. (2.3), we see that the first two terms in the sum on the left-hand side of Eq. (2.4) are zero, and we may write this equation as

$$\int_0^1 \frac{\mathrm{d}u}{\mathrm{d}x}\frac{\mathrm{d}v}{\mathrm{d}x}\,\mathrm{d}x = \int_0^1 2v(x)\,\mathrm{d}x.$$

Assuming that all derivatives and integrals we have used exist, we may now specify the *weak formulation* of our model differential equation by:

find $u(x)$ that satisfies the Dirichlet boundary conditions given by Eq. (2.2) and is such that

$$\int_0^1 \frac{\mathrm{d}u}{\mathrm{d}x}\frac{\mathrm{d}v}{\mathrm{d}x}\,\mathrm{d}x = \int_0^1 2v(x)\,\mathrm{d}x, \tag{2.5}$$

for all continuous functions $v(x)$ that satisfy $v(0) = v(1) = 0$.

The solution $u(x)$ given by Eq. (2.5) is known as the *weak solution*. The functions $v(x)$ that appear in Eq. (2.5) are known as *test functions*.

The concept of weak solutions will be revisited in the next chapter, providing some mathematical details that have been omitted above. For now, the reader need only know what has been derived above, namely that a classical solution of the differential equation and boundary conditions given by Eqs. (2.1) and (2.2) is also a weak solution that satisfies Eq. (2.5).

2.4 Elements and Nodes

We now begin to lay out some of the ingredients required by the finite element method. We have already explained that we need to partition the domain on which the differential equation is specified into smaller regions. Our model differential equation, Eq. (2.1), is defined on the domain $0 < x < 1$, and we now partition this domain into N intervals of equal length $h = \frac{1}{N}$. If we define

$$x_i = \frac{i-1}{N}, \qquad i = 1, 2, \ldots, N+1, \tag{2.6}$$

then these intervals are the regions $x_k \le x \le x_{k+1}$, for $k = 1, 2, \ldots, N$. We refer to these intervals as *elements*. The points x_i that define the element boundaries are known as *nodes*. We also define element k to be the element that occupies $x_k \le x \le x_{k+1}$, for $k = 1, 2, \ldots, N$. The elements and nodes are collectively known as the *computational mesh*, the *finite element mesh* or, more simply, the *mesh*. The nodes and elements for the case where $N = 5$ are shown in Fig. 2.1.

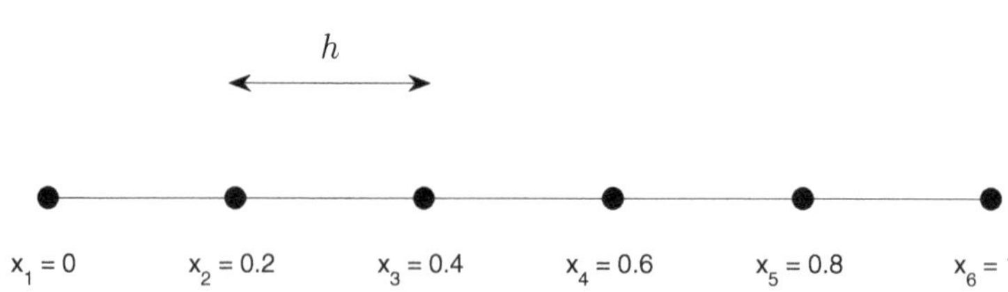

Fig. 2.1 The nodes and elements in a finite element mesh when the interval $0 < x < 1$ is partitioned into five equally sized elements

The mesh we have described above contains elements that are equally sized. We will see in the next chapter that it is straightforward to generalise the definition of the mesh so that it contains elements that are of different sizes.

2.5 Basis Functions

In this chapter, we will use the finite element method to calculate a linear approximation to the solution on each element, that is continuous across element boundaries. In the previous section, we partitioned the domain $0 < x < 1$ into N equally sized elements, each of length $h = \frac{1}{N}$. Note that this implies that

$$x_{j+1} - x_j = h, \qquad j = 1, 2, \ldots, N. \tag{2.7}$$

We now specify some functions that will be required when calculating the finite element solution. Using the property of h given by Eq. (2.7), the functions $\phi_j(x)$, for $j = 1, 2, \ldots, N + 1$, are defined by

$$\phi_1(x) = \begin{cases} (x_2 - x)/h, & x_1 \le x \le x_2, \\ 0, & \text{otherwise}, \end{cases} \tag{2.8}$$

$$\phi_j(x) = \begin{cases} (x - x_{j-1})/h, & x_{j-1} \le x \le x_j, \\ (x_{j+1} - x)/h, & x_j \le x \le x_{j+1}, \\ 0, & \text{otherwise}, \end{cases} \qquad j = 2, 3, \ldots, N, \tag{2.9}$$

$$\phi_{N+1}(x) = \begin{cases} (x - x_N)/h, & x_N \le x \le x_{N+1}, \\ 0, & \text{otherwise}. \end{cases} \tag{2.10}$$

These functions are illustrated in Fig. 2.2. For reasons that will soon become clear, we refer to these functions as *basis functions*.

The functions $\phi_j(x), j = 1, 2, \ldots, N + 1$, defined by Eqs. (2.8)–(2.10) have three properties that are useful when calculating the finite element solution: (i) they are linear functions on each element; (ii) they are continuous functions; and (iii) they satisfy, for $i = 1, 2, \ldots, N + 1$, the condition

$$\phi_j(x_i) = \begin{cases} 1, & i = j, \\ 0, & i \neq j. \end{cases} \tag{2.11}$$

These properties may easily be verified using the definition of basis functions given by Eqs. (2.8)–(2.10), and the plots of these functions in Fig. 2.2. As a consequence, the function $\phi_j(x)$ takes the value 1 at the node where $x = x_j$, and the value 0 at all other nodes. This is a very important property of basis functions, as we will see in later chapters when using more general basis functions to calculate the finite element approximation of more complex differential equations.

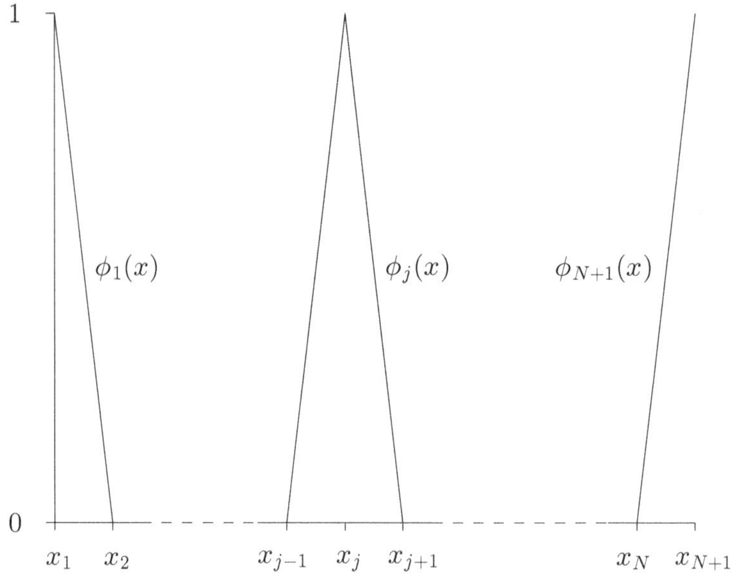

Fig. 2.2 The basis functions defined by Eqs. (2.8)–(2.10)

2.6 The Finite Element Solution

We will now use the basis functions $\phi_j(x), j = 1, 2, \ldots, N+1$, defined by Eqs. (2.8)–(2.10), to generate a finite element solution of our model differential equation that is a linear approximation to the solution on each element. We claim that we can write the finite element solution, $U(x)$, as a linear sum of these basis functions:

$$U(x) = \sum_{j=1}^{N+1} U_j \phi_j(x), \tag{2.12}$$

where the as yet unknown values U_j, $j = 1, 2, \ldots, N + 1$, are to be determined. Our claim that the finite element solution, $U(x)$, is of the form given by Eq. (2.12) is straightforward to verify. As the basis functions $\phi_j(x)$ are linear functions on each element and continuous, the sum given by Eq. (2.12) also satisfies these properties and is therefore a suitable candidate for the finite element solution. Finally we note that, on using the property of basis functions given by Eq. (2.11), we may write

$$\begin{aligned} U(x_i) &= \sum_{j=1}^{N+1} U_j \phi_j(x_i) \\ &= U_i, \end{aligned} \tag{2.13}$$

and so U_i is the finite element approximation to the solution at the node where $x = x_i$.

We now move on to describe how to determine the values U_i, $i = 1, 2, \ldots, N+1$, that complete the definition of the finite element solution given by Eq. (2.12).

2.7 Algebraic Equations Satisfied by the Finite Element Solution

Having prescribed the functional form for the finite element solution by Eq. (2.12), we now generate a system of algebraic equations satisfied by the coefficients U_j, $j = 1, 2, \ldots, N+1$. This is achieved by modifying the weak formulation of the model differential equation, given by Eq. (2.5), to specify the finite element solution $U(x)$ by:

> find $U(x)$ that satisfies the Dirichlet boundary conditions given by Eq. (2.2) and is such that
>
> $$\int_0^1 \frac{\mathrm{d}U}{\mathrm{d}x}\frac{\mathrm{d}\phi_i}{\mathrm{d}x}\,\mathrm{d}x = \int_0^1 2\phi_i(x)\,\mathrm{d}x, \tag{2.14}$$
>
> for all basis functions $\phi_i(x)$, $i = 1, 2, \ldots, N+1$, that satisfy $\phi_i(0) = \phi_i(1) = 0$.

The statement above is nothing more than the weak formulation, Eq. (2.5), restated with two minor modifications. First, we replace the weak solution $u(x)$ by the finite element solution $U(x)$. Second we restrict the test functions to be the finite set of basis functions $\phi_i(x)$ that satisfy

$$\phi_i(0) = \phi_i(1) = 0.$$

We see, from the definitions of the basis functions given by Eqs. (2.8)–(2.10) and their illustration in Fig. 2.2, that this condition on $\phi_i(x)$ is satisfied by all basis functions except $\phi_1(x)$ and $\phi_{N+1}(x)$. We therefore have $N-1$ functions $\phi_2(x), \phi_3(x), \ldots, \phi_N(x)$ that satisfy this condition.

We now use Eq. (2.14) to derive a system of algebraic equations, satisfied by the $N+1$ unknown values $U_1, U_2, \ldots, U_{N+1}$, that specify the finite element solution given by Eq. (2.12). As we have $N+1$ unknown values, we require a system of $N+1$ algebraic equations to determine these values.

The first condition in the specification of the finite element solution by Eq. (2.14) is that $U(x)$ satisfies the Dirichlet boundary conditions given by Eq. (2.2). These boundary conditions are applied at the node where $x = x_1$ and the node where $x = x_{N+1}$ and may be written as

$$U(x_1) = 0, \qquad U(x_{N+1}) = 0.$$

Using Eq. (2.13), we see that these Dirichlet boundary conditions are satisfied provided that

$$U_1 = 0, \tag{2.15}$$

$$U_{N+1} = 0, \tag{2.16}$$

and these are our first two algebraic equations.

We have already noted that the $N - 1$ basis functions ϕ_i, where $i = 2, 3, \ldots, N$, satisfy the condition on test functions given in Eq. (2.14), i.e. that $\phi_i(0) = \phi_i(1) = 0$. The remaining $N - 1$ algebraic equations result from substituting these functions into the integral equation given in Eq. (2.14). Noting that the definition of the finite element solution given by Eq. (2.12) may be differentiated to give

$$\frac{\mathrm{d}U}{\mathrm{d}x} = \sum_{j=1}^{N+1} U_j \frac{\mathrm{d}\phi_j}{\mathrm{d}x},$$

substitution of $\frac{\mathrm{d}U}{\mathrm{d}x}$ into the integral equation given by Eq. (2.14) yields

$$\int_0^1 \left(\sum_{j=1}^{N+1} U_j \frac{\mathrm{d}\phi_j}{\mathrm{d}x} \right) \frac{\mathrm{d}\phi_i}{\mathrm{d}x} \, \mathrm{d}x = \int_0^1 2\phi_i(x) \, \mathrm{d}x, \qquad i = 2, 3, \ldots, N,$$

which, after a little manipulation, becomes

$$\sum_{j=1}^{N+1} \left(\int_0^1 \frac{\mathrm{d}\phi_i}{\mathrm{d}x} \frac{\mathrm{d}\phi_j}{\mathrm{d}x} \, \mathrm{d}x \right) U_j = \int_0^1 2\phi_i(x) \, \mathrm{d}x, \qquad i = 2, 3, \ldots, N. \tag{2.17}$$

This may be written as

$$\sum_{j=1}^{N+1} A_{i,j} U_j = b_i, \qquad i = 2, 3, \ldots, N, \tag{2.18}$$

where for $i = 2, 3, \ldots, N$,

$$A_{i,j} = \int_0^1 \frac{\mathrm{d}\phi_i}{\mathrm{d}x} \frac{\mathrm{d}\phi_j}{\mathrm{d}x} \, \mathrm{d}x, \qquad j = 1, 2, 3, \ldots, N+1, \tag{2.19}$$

$$b_i = \int_0^1 2\phi_i(x) \, \mathrm{d}x. \tag{2.20}$$

We may combine Eqs. (2.15), (2.16) and (2.18) into the $(N+1) \times (N+1)$ linear system

$$AU = \mathbf{b}. \tag{2.21}$$

The entries of A and $\mathbf{b}$ in row i, where $i = 2, 3, \ldots, N$, are given by Eqs. (2.19) and (2.20). Using Eqs. (2.15) and (2.16), we see that the entries of A and $\mathbf{b}$ in rows 1 and $N+1$ are given by

$$A_{1,j} = \begin{cases} 1, & j = 1, \\ 0, & j \neq 1, \end{cases} \tag{2.22}$$

$$A_{(N+1),j} = \begin{cases} 1, & j = N+1, \\ 0, & j \neq N+1, \end{cases} \tag{2.23}$$

$$b_1 = 0, \tag{2.24}$$

$$b_{N+1} = 0. \tag{2.25}$$

The algebraic equations that comprise the linear system given by Eq. (2.21) fall into two categories. The first category is equations such as Eqs. (2.15) and (2.16) that arise from demanding that the finite element solution satisfies all Dirichlet boundary conditions. The second category is equations such as Eq. (2.18) that arise from using a suitable basis function as the test function in the integral condition given by Eq. (2.14). This categorisation of equations will be a feature of all the systems of algebraic equations that we derive in later chapters.

Having derived the linear system of algebraic equations that determines the values $U_1, U_2, \ldots, U_{N+1}$, we now introduce a flexible technique for the practical computation of the entries of the matrix A and the vector $\mathbf{b}$.

2.8 Assembling the Algebraic Equations

Most entries of the matrix A and vector $\mathbf{b}$ that appear in the linear system given by Eq. (2.21) are of the form specified by Eqs. (2.19) and (2.20) and are defined by integrals over the whole domain on which the differential equation is specified. We will now demonstrate that, from a computational implementation viewpoint, it is convenient to compute these entries by summing the contributions from individual elements. That is, for $i = 2, 3, \ldots, N$, we write Eqs. (2.19) and (2.20) as the sum of integrals over individual elements:

$$A_{i,j} = \sum_{k=1}^{N} \int_{x_k}^{x_{k+1}} \frac{\mathrm{d}\phi_i}{\mathrm{d}x} \frac{\mathrm{d}\phi_j}{\mathrm{d}x} \, \mathrm{d}x, \qquad j = 1, 2, \ldots, N+1, \tag{2.26}$$

$$b_i = \sum_{k=1}^{N} \int_{x_k}^{x_{k+1}} 2\phi_i(x) \, \mathrm{d}x. \tag{2.27}$$

Using Eq. (2.26) we see that the contribution to entry $A_{i,j}$ from integrating over the element that occupies $x_k \leq x \leq x_{k+1}$ is given by

$$\int_{x_k}^{x_{k+1}} \frac{\mathrm{d}\phi_i}{\mathrm{d}x} \frac{\mathrm{d}\phi_j}{\mathrm{d}x} \, \mathrm{d}x. \tag{2.28}$$

From the definition of the basis functions, given by Eqs. (2.8)–(2.10) and illustrated in Fig. 2.2, we see that $\phi_k(x)$ and $\phi_{k+1}(x)$ are the only basis functions that are nonzero on the element occupying $x_k \leq x \leq x_{k+1}$: all other basis functions are identically zero on this element. Hence, the integral above, over $x_k \leq x \leq x_{k+1}$, yields nonzero contributions only to the entries $A_{k,k}$, $A_{k,k+1}$, $A_{k+1,k}$, $A_{k+1,k+1}$ of A. Motivated by this observation, we calculate only these nonzero contributions when integrating over this element and store these contributions in a 2×2 matrix called $A_{\text{local}}^{(k)}$. The relationship between the location of the entries of $A_{\text{local}}^{(k)}$ and the location of the entries of A that they contribute to is given in Table 2.1.

The entries of $A_{\text{local}}^{(k)}$ are known as the *local contributions* to the matrix A from element k, and A is often referred to as the *global matrix*. We may compute A efficiently by looping over all elements, calculating only the nonzero local contributions from each element, before adding the local contributions to the appropriate entries of the global matrix A. Having calculated the entries of A that are given by Eq. (2.19), we complete the specification of the matrix by setting the remaining entries, i.e. those defined by Eqs. (2.22) and (2.23).

The entries of $\mathbf{b}$ that are given by Eq. (2.27) may be calculated in a similar manner to that used for the entries of A that are given by Eq. (2.26). The contribution to b_i, $i = 2, 3, \ldots, N$, from integrating over the element that occupies $x_k \leq x \leq x_{k+1}$ is given by

Table 2.1 The relationship between the location of the entries of $A_{\text{local}}^{(k)}$ and the location of the entries of A that they contribute to

Entry in $A_{\text{local}}^{(k)}$		Entry in A	
Row	Column	Row	Column
1	1	k	k
1	2	k	$k+1$
2	1	$k+1$	k
2	2	$k+1$	$k+1$

Table 2.2 The relationship between the location of the entries of $\mathbf{b}_{\text{local}}^{(k)}$ and the location of the entries of $\mathbf{b}$ that they contribute to

Entry in $\mathbf{b}_{\text{local}}^{(k)}$	Entry in $\mathbf{b}$
1	k
2	$k+1$

$$\int_{x_k}^{x_{k+1}} 2\phi_i(x)\,\mathrm{d}x. \tag{2.29}$$

As $\phi_k(x)$ and $\phi_{k+1}(x)$ are the only basis functions that are nonzero inside the element that occupies $x_k \leq x \leq x_{k+1}$, the only nonzero contributions to $\mathbf{b}$ from this element are to the entries b_k and b_{k+1}. We again calculate only these nonzero contributions and store them in a vector of length 2 called $\mathbf{b}_{\text{local}}^{(k)}$, known as the *local contribution* to the global vector $\mathbf{b}$. The relationship between the location of the entries of $\mathbf{b}_{\text{local}}^{(k)}$ and the location of the entries of $\mathbf{b}$ that they contribute to is given in Table 2.2. We may then calculate the global vector $\mathbf{b}$ by looping over all elements, calculating only the nonzero contributions to Eq. (2.27) from each element, and adding these contributions into $\mathbf{b}$. We then use Eqs. (2.24) and (2.25) to set the other entries.

2.8.1 *Calculating the Local Contributions*

We explained above that the entries of the global matrix A and global vector $\mathbf{b}$ that are given by Eqs. (2.19) and (2.20) should be assembled by summing the local contributions to these entries from each element. We now explain how these local contributions may be calculated.

Using the mapping between the entries of $A_{\text{local}}^{(k)}$ and the entries of A, given by Table 2.1, and the definition of the contribution to entries of A from element k, given by Eq. (2.28), we see that the entries of $A_{\text{local}}^{(k)}$ are given by

$$A_{\text{local}}^{(k)} = \begin{pmatrix} \int_{x_k}^{x_{k+1}} \frac{\mathrm{d}\phi_k}{\mathrm{d}x}\frac{\mathrm{d}\phi_k}{\mathrm{d}x}\,\mathrm{d}x & \int_{x_k}^{x_{k+1}} \frac{\mathrm{d}\phi_k}{\mathrm{d}x}\frac{\mathrm{d}\phi_{k+1}}{\mathrm{d}x}\,\mathrm{d}x \\ \int_{x_k}^{x_{k+1}} \frac{\mathrm{d}\phi_{k+1}}{\mathrm{d}x}\frac{\mathrm{d}\phi_k}{\mathrm{d}x}\,\mathrm{d}x & \int_{x_k}^{x_{k+1}} \frac{\mathrm{d}\phi_{k+1}}{\mathrm{d}x}\frac{\mathrm{d}\phi_{k+1}}{\mathrm{d}x}\,\mathrm{d}x \end{pmatrix}. \tag{2.30}$$

Similarly, using Table 2.2 and Eq. (2.29), the entries of $\mathbf{b}_{\text{local}}^{(k)}$ may be written as

$$\mathbf{b}_{\text{local}}^{(k)} = \begin{pmatrix} \int_{x_k}^{x_{k+1}} 2\phi_k(x)\,\mathrm{d}x \\ \int_{x_k}^{x_{k+1}} 2\phi_{k+1}(x)\,\mathrm{d}x \end{pmatrix}. \tag{2.31}$$

The integrals required to calculate the entries of $A_{\text{local}}^{(k)}$ and $\mathbf{b}_{\text{local}}^{(k)}$ are sufficiently simple in this case that we may evaluate them analytically. Using Eqs. (2.8)–(2.10), we see that, for $x_k \leq x \leq x_{k+1}$,

$$\phi_k(x) = \frac{x_{k+1} - x}{h},$$
$$\phi_{k+1}(x) = \frac{x - x_k}{h},$$

and so the derivatives of these functions take the constant values given by

$$\frac{\mathrm{d}\phi_k}{\mathrm{d}x} = -\frac{1}{h},$$
$$\frac{\mathrm{d}\phi_{k+1}}{\mathrm{d}x} = \frac{1}{h}.$$

Elementary integration allows us to deduce that the entries of $A_{\mathrm{local}}^{(k)}$ and $\mathbf{b}_{\mathrm{local}}^{(k)}$ are given by

$$A_{\mathrm{local}}^{(k)} = \begin{pmatrix} \frac{1}{h} & -\frac{1}{h} \\ -\frac{1}{h} & \frac{1}{h} \end{pmatrix},$$
$$\mathbf{b}_{\mathrm{local}}^{(k)} = \begin{pmatrix} h \\ h \end{pmatrix}.$$

2.8.2 Assembling the Global Matrix

We have proposed calculating the entries of the global matrix A and global vector $\mathbf{b}$ that are of the form given by Eqs. (2.19) and (2.20), by looping over all elements in the mesh, calculating the nonzero local contributions $A_{\mathrm{local}}^{(k)}$ and $\mathbf{b}_{\mathrm{local}}^{(k)}$ from each element, and using these contributions to increment the appropriate entries of A and $\mathbf{b}$, given by Tables 2.1 and 2.2.

Readers may have spotted a potential flaw with this approach for calculating A and $\mathbf{b}$. The expressions for $A_{i,j}$ and b_i given by Eqs. (2.19) and (2.20) are only valid for $i = 2, 3, \ldots, N$, with the entries for rows 1 and $N+1$ given by Eqs. (2.22)–(2.25). When calculating the local contributions from the element that occupies $x_1 \leq x \leq x_2$, we evaluate contributions to both row 1 and row 2 of A and $\mathbf{b}$. The entries of row 1 of A and $\mathbf{b}$ do not, however, fall into the pattern given by Eqs. (2.19) and (2.20), although the contribution to row 2 of both A and $\mathbf{b}$ from this element is correct. Similarly, when calculating the local contribution from the element occupying $x_N \leq x \leq x_{N+1}$ we calculate contributions to row $N+1$ of A and $\mathbf{b}$ that are not of the correct form. We could avoid adding these incorrect entries into rows 1 and $N+1$ by performing a check, before adding any local contributions into the global matrix and vector, to ensure we do not add any entries into rows 1 and $N+1$. There is nothing wrong with this approach. However, to promote clear, modular code—especially for the more complex problems encountered later in this book—it is easier to add the local contributions into rows 1 and $N+1$ and then to overwrite these erroneous entries afterwards with the correct values, given by Eqs. (2.22)–(2.25).

Having assembled the matrix A and the vector $\mathbf{b}$, most of the work required to calculate the finite element solution given by Eq. (2.12) has been done. All that remains is to solve the linear system given by Eq. (2.21) to calculate $\mathbf{U}$, before substituting the entries of $\mathbf{U}$ into Eq. (2.12). We now summarise the steps we have taken when applying the finite element method and then present an exemplar computational implementation.

2.9 A Summary of the Steps Required

The aim of this chapter has been to give an overview of the finite element method by illustrating its application to an example boundary value problem. We now give a summary of the steps required when applying the finite element method. We will see in later chapters that this summary may be used as a guide for the application of the finite element method to more general differential equations.

1. Derive the weak formulation of the boundary value problem from the specified differential equation and boundary conditions.
2. Define the finite element mesh by partitioning the domain into elements, and specifying the nodes defined by these elements.
3. Use the nodes and elements in the mesh to define suitable basis functions.
4. Write the finite element solution as a linear sum of the basis functions. Modify the weak formulation of the problem to determine the system of algebraic equations that the finite element solution satisfies. These algebraic equations will fall into one of two categories. The first category is equations that ensure that all Dirichlet boundary conditions are satisfied. The second category is equations that arise from substituting suitable test functions into an integral condition on the finite element solution.
5. Assemble the algebraic equations. Begin with the second category of equations given in step #4 above. Calculate the nonzero contributions from each element, and use these local contributions to increment the correct entries of the global matrix and vector. After this has been done, modify the equations to take account of the boundary conditions.
6. Solve the system of algebraic equations.

2.10 Computational Implementation

We will now develop a practical computational implementation of the material described in this chapter. This implementation will assemble and solve the system of algebraic equations.

The entries of the system of algebraic equations depend on the basis functions used, which in turn depend on the finite element mesh. Before we can assemble the algebraic equations, we must therefore specify the finite element mesh. The mesh we defined in Sect. 2.4 requires only specification of the number of elements, N. This is sufficient to define the location of the nodes $x_1, x_2, \ldots, x_{N+1}$, and the element length h. Further, these nodes allow us to specify the basis functions through the material given in Sect. 2.5. We are then in a position to assemble the algebraic equations. We will therefore implement the material described in this chapter by writing a function that accepts a positive integer, N, that represents the number of elements, as an input, and returns the vector $\mathbf{x}$ (containing the nodes in the mesh) and the vector $\mathbf{U}$ (containing the finite element solution at each node).

A MATLAB script for the implementation described above is given in Listing 2.1. This function may be called by saving the file as `Chap2_CalculateModelFem-Solution.m` and typing, in a MATLAB session,

```
[x, U] = Chap2_CalculateModelFemSolution(40);
```

to run with $N = 40$ elements. Variable names in this script have been chosen to correspond to the mathematical notation used in this chapter. This function begins by initialising entries of the global matrix A and global vector $\mathbf{b}$ to zero in lines 4 and 5, before generating a vector containing the equally spaced nodes in the mesh in line 8. We then loop over all elements in lines 12–23, calculating the nonzero local contributions to A and $\mathbf{b}$ given by Eqs. (2.30) and (2.31), before incrementing the appropriate entries of the global linear system, given by Tables 2.1 and 2.2. MATLAB users will be aware that the expression `A(k:k+1, k:k+1)` that appears in line 21 is shorthand for the submatrix given by

$$\begin{pmatrix} A_{k,k} & A_{k,k+1} \\ A_{k+1,k} & A_{k+1,k+1} \end{pmatrix},$$

and that the expression `b(k:k+1)` that appears in line 22 represents the subvector given by

$$\begin{pmatrix} b_k \\ b_{k+1} \end{pmatrix}.$$

Note that the loop over elements introduces incorrect entries into rows 1 and $N+1$ of A and $\mathbf{b}$, for the reasons explained in Sect. 2.8.2. These entries are overwritten with the correct values in lines 26–31. Note that line 26 uses MATLAB notation to set all entries of row 1 to zero. We then solve the linear system using MATLAB's backslash operator in line 34. This is a suitable method for solving linear systems of the modest size generated here. However, iterative techniques may be more suitable for solving the larger systems of algebraic equations that arise from the application of the finite element method to partial differential equations in later chapters. A summary of these linear algebra techniques is given in Appendix A. Finally, in lines 37–39, we plot the finite element solution that has been calculated, which is a linear approximation to the solution on each element. This finite element solution may be plotted by drawing a straight line between the finite element solution at adjacent nodes: this is exactly what is plotted by line 37 of the listing.

Listing 2.1 `Chap2_CalculateModelFemSolution.m`, a MATLAB function for calculating the finite element solution of the model problem.

```
function [x, U] = Chap2_CalculateModelFemSolution(N)

% Initialise A and b to zero
A = zeros(N+1, N+1);
b = zeros(N+1, 1);

% Generate N+1 nodes, equally spaced between 0 and 1
x = linspace(0, 1, N+1);

% Loop over elements calculating local contributions
% and incrementing the global linear system
for k=1:N
    % Calculate element length
    h = 1/N;

    % Calculate local contributions
    A_local = [1/h, -1/h; -1/h, 1/h];
    b_local = [h; h];

    % Increment global system
    A(k:k+1, k:k+1) = A(k:k+1, k:k+1) + A_local;
    b(k:k+1) = b(k:k+1) + b_local;
end

% Set Dirichlet boundary conditions
A(1,:) = 0;
A(1,1) = 1;
b(1) = 0;
A(N+1,:) = 0;
A(N+1,N+1) = 1;
b(N+1) = 0;

% Solve linear system and plot FE solution
U = A\b;

% Plot finite element solution
plot(x, U, '-o')
xlabel('x')
ylabel('U')
```

2.11 Evaluating the Finite Element Solution at a Given Point

We have now developed a computational implementation of the finite element method that computes the quantities $U_1, U_2, \ldots, U_{N+1}$ that appear in the finite element solution given by Eq. (2.12). We will often need to evaluate the solution at some given point x, that is, evaluate $U(x)$. If x is a node in the mesh, i.e. $x_i = x$ for some i, we may then use Eq. (2.13) to write

$$U(x) = U_i.$$

If x is not a node in the mesh, then it will lie in element k for some k, where $x_k < x < x_{k+1}$. We first locate the element k that x lies in. Noting, as we did earlier in this chapter, that only the basis functions $\phi_k(x)$ and $\phi_{k+1}(x)$ are nonzero on element k, the finite element solution given by Eq. (2.12) may be written, for $x_k < x < x_{k+1}$, as

$$U(x) = U_k\phi_k(x) + U_{k+1}\phi_{k+1}(x).$$

We then use the definition of the basis functions $\phi_k(x)$ and $\phi_{k+1}(x)$, given by Eqs. (2.8)–(2.10), to deduce that

$$U(x) = \frac{x_{k+1} - x}{h}U_k + \frac{x - x_k}{h}U_{k+1}.$$

2.12 Is the Finite Element Solution Correct?

It is always worth checking the solution to any mathematical problem, whether the solution has been obtained by analytic "pen and paper" techniques or as the output from a computer program. Easily made typographical errors, such as a missed factor of 2, will at best lead to inaccuracies in the solution. There is, however, the potential for completely nonsensical output, for example a physical quantity being assigned a value that is calculated from taking the square root of a negative number. It is, therefore, useful to perform some verification of the computation of a finite element solution. An exhaustive verification, such as comparing the finite element solution of a differential equation with the true solution of the equation, is rarely possible. After all, if we knew the true solution of a differential equation, we would be unlikely to go to the effort of computing the finite element solution. Nevertheless, some validation is possible. Below we list a few simple tests that can usually be carried out.

- The simplest test that can be carried out is to plot the finite element solution to check that the solution looks realistic and takes values that are reasonable—a computed chemical concentration that takes a negative value, for example, should set alarm bells ringing.
- Another simple check that should be carried out is to compare finite element solutions that have been calculated using different numbers of elements. The finite element solution should become more accurate, and approach a limit as the number of elements is successively increased.
- Finally, analytic solutions may exist for some choices of parameter values that appear in the differential equation, or for simplifications of the differential equa-

tion. The finite element solution should be computed in these cases and compared to the analytic solution.

2.13 Exercises

For these exercises, you will need to download the MATLAB script given in Listing 2.1. Remember that the true solution to the differential equation on which this script is based is $u(x) = x(1-x)$.

2.1 In this exercise, we will modify the MATLAB script given in Listing 2.1 so that it returns the finite element solution at the point $x = 0.45$.

(a) Modify the MATLAB script so that it returns the finite element solution at $x = 0.45$ when this point is a node of the mesh. Use the true solution, and a mesh with $N = 100$ elements, to verify that your answer is correct.
(b) Use the material given in Sect. 2.11 to modify the MATLAB script to evaluate the finite element solution at $x = 0.45$ when this point is not a node of the mesh. Use the true solution, and a mesh with $N = 101$ elements, to verify that your answer is correct.
(c) Combine your answers to parts (a) and (b) so that the MATLAB script first determines whether or not $x = 0.45$ is a node of the mesh, before returning the finite element solution at that point.

2.2 We denote the error between the finite element solution and the true solution by $E(x) = U(x) - u(x)$. The L^2 norm of the error of the finite element solution, $\|E\|_{L^2}$, is defined to be

$$\|E\|_{L^2} = \sqrt{\int_0^1 (U(x) - u(x))^2 \, \mathrm{d}x},$$

which may be written as the sum of the contribution from individual elements:

$$\|E\|_{L^2} = \sqrt{\sum_{k=1}^{N} \int_{x_k}^{x_{k+1}} (U(x) - u(x))^2 \, \mathrm{d}x}.$$

(a) In Sect. 2.11, we saw that, for $x_k < x < x_{k+1}$, the finite element solution is given by

$$U(x) = \frac{x_{k+1} - x}{h} U_k + \frac{x - x_k}{h} U_{k+1}.$$

Use this expression for $U(x)$ to explain why, for the model problem used in this chapter, we may write

$$\|E\|_{L^2}^2 = \sum_{k=1}^{N} \int_{x_k}^{x_{k+1}} \left(\frac{x_{k+1} - x}{h} U_k + \frac{x - x_k}{h} U_{k+1} - x(1-x) \right)^2 \, \mathrm{d}x.$$

(b) Evaluate, by direct integration, the quantity

$$\int_{x_k}^{x_{k+1}} \left(\frac{x_{k+1} - x}{h} U_k + \frac{x - x_k}{h} U_{k+1} - x(1-x) \right)^2 \, \mathrm{d}x,$$

and write your answer as a function of U_k and U_{k+1}.

(c) Using your answers to parts (a) and (b), modify the MATLAB script given in Listing 2.1 so that it outputs the value of $\|E\|_{L^2}$ for a given number of elements, N.

(d) Define the element length by $h = \frac{1}{N}$ where N is the number of elements. You may assume that, as $h \to 0$, $\|E\|_{L^2}$ is related to the element length h by

$$\|E\|_{L^2} = Ch^p,$$

for some constants C and p. Taking the logarithm of both sides of this expression gives

$$\log\left(\|E\|_{L^2}\right) = \log C + p \log h,$$

and so plotting $\log\left(\|E\|_{L^2}\right)$ against $\log h$ will give a straight line of gradient p as $h \to 0$. By plotting $\log\left(\|E\|_{L^2}\right)$ against $\log h$ for several values of h, estimate the constant p.

2.3 In Sect. 2.11, we saw that, for $x_k < x < x_{k+1}$, the finite element solution is given by

$$U(x) = \frac{x_{k+1} - x}{h} U_k + \frac{x - x_k}{h} U_{k+1}.$$

(a) Write down the derivative of the finite element solution, $\frac{\mathrm{d}U}{\mathrm{d}x}$, on element k.

(b) Modify the MATLAB script given in Listing 2.1 so that it plots the derivative of the solution. Note that this derivative is a function that may be discontinuous across element boundaries.

2.4 A boundary value problem is defined by

$$-3\frac{\mathrm{d}^2 u}{\mathrm{d}x^2} = 2, \qquad 0 < x < 1,$$

together with boundary conditions

$$u(0) = 1, \qquad u(1) = 2.$$

(a) Write down the analytic solution to this boundary value problem.
(b) By working through the steps required to calculate the finite element solution, summarised in Sect. 2.9, modify the MATLAB script given in Listing 2.1 to calculate the finite element approximation of the boundary value problem above. Compare your finite element solution to the true solution.

Chapter 3
Linear Boundary Value Problems

In the previous chapter, we demonstrated how to calculate the finite element solution of a very simple differential equation. The intention of this simple example was to give the reader an overview of the steps required when applying the finite element method. As a consequence, for the sake of clarity, some mathematical rigour was neglected. In this chapter, we introduce more rigour into our description of the finite element method and demonstrate how to calculate the finite element solution of general linear, second-order, boundary value problems.

3.1 A General Boundary Value Problem

A general linear, second-order, boundary value problem, defined on the interval $a < x < b$, may be written in the form

$$-\frac{\mathrm{d}}{\mathrm{d}x}\left(p(x)\frac{\mathrm{d}u}{\mathrm{d}x}\right) + q(x)\frac{\mathrm{d}u}{\mathrm{d}x} + r(x)u = f(x), \tag{3.1}$$

for given functions $p(x)$, $q(x)$, $r(x)$ and $f(x)$, where the function $p(x)$ satisfies the condition that $p(x) \neq 0$ for $a < x < b$. A differential equation for $u(x)$ is described as *linear* provided the differential equation is linear in $u(x)$ and its derivatives. The equation above will then still classified as a linear differential equation even if the functions $p(x)$, $q(x)$, $r(x)$ and $f(x)$ are nonlinear functions of x.

To fully specify the boundary value problem, we require one boundary condition at $x = a$ and one boundary condition at $x = b$. At this point, we consider only two types of boundary conditions: *Dirichlet boundary conditions*, where the value of the function $u(x)$ is specified on the boundary, and *Neumann boundary conditions*, where the derivative of the function $u(x)$ is specified on the boundary. We will consider more general boundary conditions in Sect. 3.8. For the purpose of illustrating the application of the finite element method, we will specify one Dirichlet boundary condition and one Neumann boundary condition. This will allow us to demonstrate

J. Whiteley, *Finite Element Methods*, Mathematical Engineering,
DOI 10.1007/978-3-319-49971-0_3

how both categories of boundary condition are handled by the finite element method. It should then be clear how to adapt the finite element method for any suitable combination of these boundary conditions. We impose a Dirichlet boundary condition at $x = a$:

$$u = u_a, \qquad x = a, \tag{3.2}$$

for some given constant u_a and a Neumann boundary condition at $x = b$:

$$p(x)\frac{\mathrm{d}u}{\mathrm{d}x} = \eta_b, \qquad x = b, \tag{3.3}$$

where η_b is another given constant. Note the presence of $p(x)$ in both Eqs. (3.1) and (3.3): the left-hand side of Eq. (3.3) has the same pattern as the contents of the brackets on the left-hand side of Eq. (3.1). A Neumann boundary condition written in this form is known as a *natural boundary condition*. The reason for this terminology will become clear shortly when we derive the weak solution of the differential equation given by Eq. (3.1), subject to the boundary conditions given by Eqs. (3.2) and (3.3).

We have written the Neumann boundary condition, Eq. (3.3), as a natural Neumann boundary condition. It is straightforward to write other Neumann boundary conditions, such as

$$\frac{\mathrm{d}u}{\mathrm{d}x} = \hat{\eta}_b, \qquad x = b, \tag{3.4}$$

as natural Neumann boundary conditions; we simply multiply the boundary condition given by Eq. (3.4) by $p(x)$, which we assumed earlier was nonzero. Writing $\eta_b = p(b)\hat{\eta}_b$ then yields a natural Neumann boundary condition in the form of Eq. (3.3).

3.1.1 A Note on the Existence and Uniqueness of Solutions

It is important to be aware that there is no guarantee that a solution exists to a general boundary value problem. Furthermore, if a solution does exist, then it may not be unique. For example, suppose we attempt to solve the differential equation defined by, for $0 < x < \pi$:

$$\frac{\mathrm{d}^2u}{\mathrm{d}x^2} + u = 0, \tag{3.5}$$

subject to the boundary conditions

$$u(0) = 0, \qquad u(\pi) = 1. \tag{3.6}$$

The general solution of Eq. (3.5) is $u(x) = K_1 \sin x + K_2 \cos x$, for constants K_1 and K_2 that we would expect to determine from the boundary conditions given by Eq. (3.6). However, it is not possible to find values of K_1 and K_2 that satisfy both of these boundary conditions, and we deduce that no solution exists to this boundary value problem.

Suppose, instead, that the boundary conditions are given by

$$u(0) = 0, \qquad u(\pi) = 0. \tag{3.7}$$

The solution of Eq. (3.5), subject to the boundary conditions given by Eq. (3.7), is $u(x) = K_1 \sin x$ for *any* constant K_1. Hence, the solution is not unique. A full discussion of the existence and uniqueness of solutions to boundary value problems is beyond the scope of this book. Throughout this book, you may assume, unless stated otherwise, that a unique solution exists to any differential equation considered.

3.2 The Weak Formulation

The discussion of the weak formulation of a boundary value problem in Sect. 2.3, although entirely adequate for the purposes of Chap. 2, lacked sufficient detail to allow the reader to apply the finite element method to a general boundary value problem such as that presented in Sect. 3.1. In particular, the model problem considered in Chap. 2 included only the very simple Dirichlet boundary conditions that $u(x) = 0$ on both boundaries. As promised, we now provide more detail, allowing more general boundary conditions to be handled. Our treatment of the weak formulation will also be more rigorous than that presented in Chap. 2.

3.2.1 *Some Definitions*

The weak formulation of the example differential equation derived in Sect. 2.3 assumed the existence of several derivatives and integrals. We now provide some notation that aids us in deriving the weak formulation of differential equations in a more rigorous manner. Suppose that a function $v(x)$ is defined on the interval $a \leq x \leq b$. We say that $v(x)$ is *square integrable* on this interval if the integral

$$\int_a^b (v(x))^2 \, \mathrm{d}x$$

exists: in practice, this integral is said to exist if it takes a finite value. We define $L^2(a, b)$ to be a set of all functions $v(x)$ that are square integrable on the interval $a \leq x \leq b$ and write $L^2(a, b)$ in set notation as

$$L^2(a,b) = \left\{ \text{functions } v(x) \text{ such that the integral } \int_a^b (v(x))^2 \, \mathrm{d}x \text{ exists} \right\}. \tag{3.8}$$

We now define the Sobolev space of order 1, denoted by $H^1(a,b)$, to be the set of functions $v(x)$ that: (i) are defined and continuous on the interval $a \leq x \leq b$; (ii) are square integrable on $a \leq x \leq b$; and (iii) possess a first derivative that is square integrable on $a \leq x \leq b$. In set notation,

$$H^1(a,b) = \left\{ \text{functions } v(x) \text{ such that } v(x) \text{ is continuous, and the integrals } \int_a^b (v(x))^2 \, \mathrm{d}x \text{ and } \int_a^b \left(\frac{\mathrm{d}v}{\mathrm{d}x}\right)^2 \mathrm{d}x \text{ exist} \right\}. \tag{3.9}$$

Suppose that a differential equation, together with suitable boundary conditions, has been specified. Using these boundary conditions, we specify two subsets of $H^1(a,b)$ that will be required later:

$$H^1_E(a,b) = \left\{ \text{functions } v(x) \text{ in } H^1(a,b) \text{ such that } v(x) \text{ satisfies all Dirichlet boundary conditions} \right\}, \tag{3.10}$$

$$H^1_0(a,b) = \left\{ \text{functions } v(x) \text{ in } H^1(a,b) \text{ such that } v(x) = 0 \text{ at all points } x \text{ where Dirichlet boundary conditions are specified} \right\}. \tag{3.11}$$

The use of these sets of functions when deriving the weak formulation of a differential equation ensures the existence of all integrals that are used.

3.2.2 Deriving the Weak Formulation

The definitions of the Sobolev space of order 1, and subsets of this space, laid out in Eqs. (3.9)–(3.11), may be used to aid the derivation of the weak formulation of the differential equation given by Eq. (3.1), subject to the boundary conditions given by Eqs. (3.2) and (3.3). We begin by multiplying the differential equation, Eq. (3.1), by any test function $v(x) \in H^1_0(a,b)$, and integrating the resulting product over the interval $a < x < b$:

$$\int_a^b -\frac{\mathrm{d}}{\mathrm{d}x}\left(p\frac{\mathrm{d}u}{\mathrm{d}x}\right) v + q\frac{\mathrm{d}u}{\mathrm{d}x} v + ruv \, \mathrm{d}x = \int_a^b fv \, \mathrm{d}x.$$

We now integrate the first term on the left-hand side by parts to give

$$\left[-p\frac{\mathrm{d}u}{\mathrm{d}x}v\right]_a^b + \int_a^b p\frac{\mathrm{d}u}{\mathrm{d}x}\frac{\mathrm{d}v}{\mathrm{d}x} + q\frac{\mathrm{d}u}{\mathrm{d}x}v + ruv\,\mathrm{d}x = \int_a^b fv\,\mathrm{d}x. \tag{3.12}$$

We have chosen $v(x) \in H_0^1(a,b)$ and so, by the definition of this set given in Eq. (3.11), we have $v(x) = 0$ at any value of x where a Dirichlet boundary condition is applied. As a Dirichlet boundary condition, given by Eq. (3.2), is applied at $x = a$, we have $v(a) = 0$. Substituting both this Dirichlet boundary condition, and the Neumann boundary condition given by Eq. (3.3) into Eq. (3.12) gives, after a little rearrangement:

$$\int_a^b p\frac{\mathrm{d}u}{\mathrm{d}x}\frac{\mathrm{d}v}{\mathrm{d}x} + q\frac{\mathrm{d}u}{\mathrm{d}x}v + ruv\,\mathrm{d}x = \int_a^b fv\,\mathrm{d}x + \eta_b v(b). \tag{3.13}$$

Note that writing the Neumann boundary condition, Eq. (3.3), as a natural boundary condition allows the contribution to Eq. (3.13) from the Neumann boundary condition to be independent of the function $p(x)$.

In deriving Eq. (3.13), we have used the differential equation, Eq. (3.1), and the Neumann boundary condition, Eq. (3.3). We have not yet used the Dirichlet boundary condition given by Eq. (3.2), other than to specify the set $H_0^1(a,b)$. We satisfy the Dirichlet boundary condition by demanding that the weak solution lies in the set $H_E^1(a,b)$ defined by Eq. (3.10), as all functions in this set satisfy this Dirichlet boundary condition.

Using Eq. (3.13), we may now define the weak solution of Eqs. (3.1)–(3.3) by:

find $u(x) \in H_E^1(a,b)$ such that

$$\int_a^b p\frac{\mathrm{d}u}{\mathrm{d}x}\frac{\mathrm{d}v}{\mathrm{d}x} + q\frac{\mathrm{d}u}{\mathrm{d}x}v + ruv\,\mathrm{d}x = \int_a^b fv\,\mathrm{d}x + \eta_b v(b), \tag{3.14}$$

for all $v(x) \in H_0^1(a,b)$.

The derivation above has illustrated one key feature of the weak formulation of a differential equation. Dirichlet boundary conditions are satisfied through specifying the set in which $u(x)$ lies, i.e. the set $H_E^1(a,b)$. Neumann boundary conditions are handled through incorporating these boundary conditions into Eq. (3.14), the equation that is satisfied by the weak solution for all suitable test functions. We will see in the next section that this remark also applies to the finite element solution.

3.3 The Finite Element Solution

We now proceed as in Chap. 2 by specifying the computational mesh and writing the finite element solution as a linear sum of basis functions defined on this mesh. We then use a modification of the weak formulation of the differential equation to identify the system of algebraic equations that the coefficients of the basis functions in this linear sum satisfy.

3.3.1 The Mesh

In Chap. 2, we used a mesh with equally spaced nodes, often known as a *uniform mesh*, when computing the finite element solution. A consequence of using equally spaced nodes is that all elements are the same size. We now relax this condition and allow computational meshes where the nodes are not equally spaced across the computational domain $a < x < b$.

As before, the computational mesh will consist of N elements and $N + 1$ nodes. The nodes are again denoted by $x_1, x_2, \ldots, x_{N+1}$. We demand that these nodes satisfy three conditions: (i) $x_1 = a$, (ii) $x_{N+1} = b$ and (iii) $x_{k+1} > x_k$ for $k = 1, 2, \ldots, N$.

Defining element k to occupy the region $x_k \leq x \leq x_{k+1}$, we see that nodes satisfying the conditions above generate elements that cover the whole of the domain $a < x < b$ without overlapping, and all have positive length. The mesh containing these nodes and elements is therefore a suitable partitioning of the domain.

3.3.2 Basis Functions

We will now define basis functions that allow us to calculate a finite element solution that is a linear approximation to the solution of a differential equation on each element. These basis functions will be defined using two different formulations, which we will show are equivalent. The first formulation defines the basis functions by modifying the basis functions used in Chap. 2, so that they may be applied on the nonuniform mesh described in Sect. 3.3.1. We then present an alternative formulation for defining these basis functions. Both methods for defining the basis functions generate identical basis functions. We will see, however, that the second technique that we present is more flexible. Later in this chapter, we will see that the linear system that arises from the implementation of the finite element method includes some entries that are defined by integrals that are difficult, or impossible, to evaluate analytically. The second method for defining basis functions will aid us in approximating these integrals using numerical techniques. Further, we will see in later chapters that the second method for defining basis functions allows the material presented in this chapter to be extended, relatively easily, to both higher order basis functions and the application of the finite element method to partial differential equations. We now describe both of these formulations.

3.3.2.1 Generalising the Basis Functions Used in Chap. 2

The basis functions defined by Eqs. (2.8)–(2.10) assume that each element is of a constant length h. Noting that element k is of length $x_{k+1} - x_k$, for $k = 1, 2, \ldots, N$, it is straightforward to modify these basis functions for use on the nonuniform mesh described in Sect. 3.3.1, by replacing the constant h in Eqs. (2.8)–(2.10) by the appropriate element length. We may then define the basis functions $\phi_j(x)$, for $j = 1, 2, \ldots, N+1$, on the nonuniform mesh described in Sect. 3.3.1 by

$$\phi_1(x) = \begin{cases} \frac{x_2 - x}{x_2 - x_1}, & x_1 \le x \le x_2, \\ 0, & \text{otherwise}, \end{cases} \tag{3.15}$$

$$\phi_j(x) = \begin{cases} \frac{x - x_{j-1}}{x_j - x_{j-1}}, & x_{j-1} \le x \le x_j, \\ \frac{x_{j+1} - x}{x_{j+1} - x_j}, & x_j \le x \le x_{j+1}, \\ 0, & \text{otherwise}, \end{cases} \qquad j = 2, 3, \ldots, N, \tag{3.16}$$

$$\phi_{N+1}(x) = \begin{cases} \frac{x - x_N}{x_{N+1} - x_N}, & x_N \le x \le x_{N+1}, \\ 0, & \text{otherwise}. \end{cases} \tag{3.17}$$

Should each element have the same length h, the basis functions defined above are identical to the basis functions defined by Eqs. (2.8)–(2.10).

When deriving the basis functions in Sect. 2.5, we noted that the basis functions had three useful properties: (i) they are linear functions on each element; (ii) they are continuous functions; and (iii) they satisfy, for $j = 1, 2, \ldots, N+1$, the condition

$$\phi_j(x_i) = \begin{cases} 1, & i = j, \\ 0, & i \ne j. \end{cases} \tag{3.18}$$

It is straightforward to verify that these properties are also true for the basis functions defined above. Writing the finite element solution as

$$U(x) = \sum_{j=1}^{N+1} U_j \phi_j(x), \tag{3.19}$$

where $\phi_j(x)$, $j = 1, 2, \ldots, N+1$, are defined by Eqs. (3.15)–(3.17), we see that, as required, our finite element solution is a linear function on each element that is continuous across element boundaries. Furthermore, Eqs. (3.18) and (3.19) allow us to write

$$U(x_i) = U_i, \qquad i = 1, 2, \ldots, N+1, \tag{3.20}$$

and so U_i is, as before, the finite element solution at the node where $x = x_i$.

3.3.2.2 A More General Method for Defining the Basis Functions

As promised earlier, we will now present a more general definition of the basis functions. The global basis functions are now defined through: (i) a mapping between a general element in the mesh and a canonical element, and (ii) local basis functions defined on the canonical element.

We may map between element k of the mesh described in Sect. 3.3.1, and the unit interval $0 \leq X \leq 1$, by using the change of variable

$$x = x_k + (x_{k+1} - x_k)X. \tag{3.21}$$

The unit interval $0 \leq X \leq 1$ is known as the *canonical element*. We now define local basis functions $\phi_{\text{local},1}(X)$, $\phi_{\text{local},2}(X)$ on the canonical element by

$$\phi_{\text{local},1}(X) = 1 - X, \qquad 0 \leq X \leq 1, \tag{3.22}$$
$$\phi_{\text{local},2}(X) = X, \qquad 0 \leq X \leq 1. \tag{3.23}$$

Global basis functions ϕ_j, where $j = 1, 2, \ldots, N+1$, may then be defined on element k, in terms of the change of variable given by Eq. (3.21), and the local basis functions $\phi_{\text{local},1}(X)$, $\phi_{\text{local},2}(X)$ defined by Eqs. (3.22) and (3.23). On element k, we set

$$\phi_k(X) = \phi_{\text{local},1}(X), \qquad 0 \leq X \leq 1, \tag{3.24}$$
$$\phi_{k+1}(X) = \phi_{\text{local},2}(X), \qquad 0 \leq X \leq 1, \tag{3.25}$$
$$\phi_j(X) = 0, \qquad 0 \leq X \leq 1, \qquad j \neq k, j \neq k+1. \tag{3.26}$$

Equations (3.21)–(3.26) then define the global basis functions ϕ_j, for $j = 1, 2, \ldots, N+1$, for every element $k = 1, 2, \ldots, N$.

We now demonstrate that the basis functions defined by Eqs. (3.21)–(3.26) are identical to those defined by Eqs. (3.15)–(3.17). Using the change of variable given by Eq. (3.21), and the definition of the local basis functions given by Eqs. (3.22) and (3.23), we may write $\phi_{\text{local},1}$, $\phi_{\text{local},2}$ as functions of x on element k:

$$\phi^{(k)}_{\text{local},1}(x) = \frac{x_{k+1} - x}{x_{k+1} - x_k}, \qquad x_k \leq x \leq x_{k+1}, \tag{3.27}$$
$$\phi^{(k)}_{\text{local},2}(x) = \frac{x - x_k}{x_{k+1} - x_k}, \qquad x_k \leq x \leq x_{k+1}. \tag{3.28}$$

We introduce the superscript (k) to the functions $\phi^{(k)}_{\text{local},1}(x)$, $\phi^{(k)}_{\text{local},2}(x)$ when writing these functions as a function of x, to emphasise that they depend on the element through the change of variable given by Eq. (3.21). We place this superscript in brackets to avoid any confusion with these functions being raised to the power k.

We may now use Eqs. (3.27) and (3.28) to write the global basis functions given by Eqs. (3.24)–(3.26) as functions of x on element k:

$$\phi_k(x) = \frac{x_{k+1} - x}{x_{k+1} - x_k}, \qquad x_k \le x \le x_{k+1}, \qquad (3.29)$$

$$\phi_{k+1}(x) = \frac{x - x_k}{x_{k+1} - x_k}, \qquad x_k \le x \le x_{k+1}, \qquad (3.30)$$

$$\phi_j(x) = 0, \qquad x_k \le x \le x_{k+1}, \qquad j \ne k,\ j \ne k+1. \qquad (3.31)$$

These basis functions are plotted as a function of x on element k in Fig. 3.1.

We see that the basis functions defined on any element k by Eqs. (3.21)–(3.26), and plotted in Fig. 3.1, are identical to the basis functions defined by Eqs. (3.15)–(3.17) when restricted to element k. This is true for any element k, and so the basis functions defined by Eqs. (3.21)–(3.26) are identical to those defined by Eqs. (3.15)–(3.17). Properties of basis functions derived in Sect. 3.3.2.1, such as those given by Eqs. (3.18) and (3.20), therefore hold for the basis functions defined in this section.

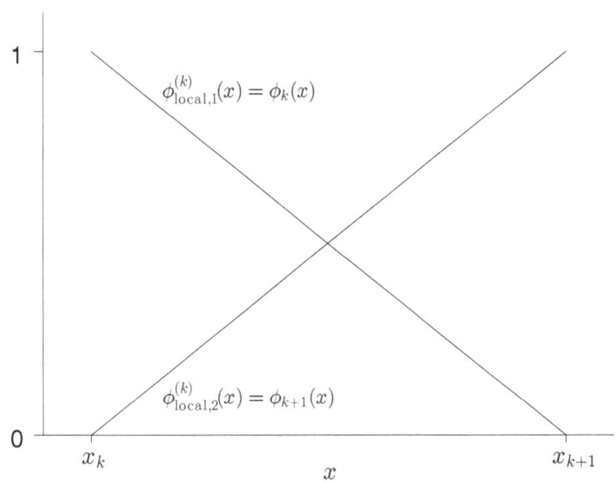

Fig. 3.1 The basis functions $\phi_k(x)$ and $\phi_{k+1}(x)$ defined by Eqs. (3.29) and (3.30)

3.3.3 Sets of Functions

We now identify some sets of potential finite element solutions that mirror the Sobolev spaces that were used when deriving the weak formulation in Sect. 3.2. First, we specify the set of all potential finite element solutions that do not take account of any boundary conditions, i.e. the set of continuous functions, that are linear functions when restricted to a given element. This set, denoted by S^h, is defined by

$$S^h = \left\{ \text{functions } V(x) \text{ given by } V(x) = \sum_{j=1}^{N+1} V_j \phi_j(x) \text{ where } \phi_j,\ j = 1, 2, \ldots, N+1, \right.$$
$$\left. \text{are suitably defined basis functions} \right\}. \tag{3.32}$$

Subsets of S^h are defined as follows:

$$S_E^h = \left\{ \text{functions } V(x) \in S^h \text{ such that} V(x) \text{ satisfies all Dirichlet boundary} \right.$$
$$\left. \text{conditions} \right\}, \tag{3.33}$$

$$S_0^h = \left\{ \text{functions } V(x) \in S^h \text{ such that } V(x) = 0 \text{ at all points } x \text{ where Dirichlet} \right.$$
$$\left. \text{boundary conditions are specified} \right\}. \tag{3.34}$$

3.3.4 The Finite Element Solution

Having defined the set S^h, and the subsets S_E^h and S_0^h of this set, in Sect. 3.3.3, we are now in a position to define the finite element solution. As in Chap. 2, this is done by modifying the definition of the weak solution, given by Eq. (3.14). The sets of functions $H_E^1(0, 1)$ and $H_0^1(0, 1)$ that appear in the weak solution are replaced by the sets S_E^h and S_0^h, allowing us to specify the finite element solution by:

find $U(x) \in S_E^h$ such that

$$\int_a^b p \frac{\mathrm{d}U}{\mathrm{d}x} \frac{\mathrm{d}\phi_i}{\mathrm{d}x} + q \frac{\mathrm{d}U}{\mathrm{d}x} \phi_i + rU\phi_i \,\mathrm{d}x = \int_a^b f\phi_i \,\mathrm{d}x + \eta_b \phi_i(b), \tag{3.35}$$

for all $\phi_i(x) \in S_0^h$.

3.3.5 The System of Algebraic Equations

We may derive the system of algebraic equations satisfied by $U_1, U_2, \ldots, U_{N+1}$ in a similar way to that described in Sect. 2.7. We first write the finite element solution as

$$U(x) = \sum_{j=1}^{N+1} U_j \phi_j(x), \tag{3.36}$$

where $\phi_j(x)$, $j = 1, 2, \ldots, N+1$, are suitably defined basis functions. Earlier, in Sects. 3.3.2.1 and 3.3.2.2, we gave two definitions of basis functions that we demonstrated are identical. As such, for the purposes of this section, either definition suffices. We note that $U(x)$ is a member of the set S^h.

The finite element solution given by Eq. (3.36) contains $N+1$ unknown values, $U_1, U_2, \ldots, U_{N+1}$. We therefore need $N+1$ algebraic equations to determine these unknown values. As in Chap. 2, these equations will be drawn from two sources: equations that ensure that $U(x)$ satisfies the Dirichlet boundary conditions and equations that arise from substituting suitable basis functions $\phi_i(x)$ into the integral condition given by Eq. (3.35).

The finite element solution, defined by Eq. (3.35), requires that $U(x) \in S_E^h$. We have already noted that $U(x)$, given by Eq. (3.36), lies in the set S^h. To ensure that $U(x)$ lies in the subset S_E^h of the set S^h, we need $U(x)$ to satisfy any prescribed Dirichlet boundary conditions. We have one Dirichlet boundary condition, given by Eq. (3.2), and this boundary condition applies at $x = a$: that is, at the node where $x = x_1$. Applying Eq. (3.20) at the this node gives

$$U(x_1) = U_1, \tag{3.37}$$

and so this Dirichlet boundary condition is satisfied if

$$U_1 = u_a. \tag{3.38}$$

The condition that $U(x)$ satisfies the Dirichlet boundary condition has allowed us to derive one algebraic equation, Eq. (3.38). We therefore require a further N algebraic equations to determine the $N+1$ unknown values. Using Eq. (3.18), we see that $\phi_i(x_1) = 0$ for $i = 2, 3, \ldots, N+1$. Hence, $\phi_i(x) \in S_0^h$ for $i = 2, 3, \ldots, N+1$, as these functions take the value zero at the value of x where Dirichlet boundary conditions are applied. Substituting $\phi_i(x)$, for $i = 2, 3, \ldots, N+1$, into Eq. (3.35) will then generate a further N algebraic equations, as required. Noting that Eq. (3.36) implies that

$$\frac{\mathrm{d}U}{\mathrm{d}x} = \sum_{j=1}^{N+1} U_j \frac{\mathrm{d}\phi_j}{\mathrm{d}x}, \tag{3.39}$$

we have, on substituting $\phi_i(x)$, $i = 2, 3, \ldots, N+1$, into Eq. (3.35) and rearranging, the N algebraic equations

$$\sum_{j=1}^{N+1} \left(\int_a^b p \frac{\mathrm{d}\phi_j}{\mathrm{d}x} \frac{\mathrm{d}\phi_i}{\mathrm{d}x} + q \frac{\mathrm{d}\phi_j}{\mathrm{d}x} \phi_i + r\phi_j\phi_i \, \mathrm{d}x \right) U_j = \int_a^b f\phi_i \, \mathrm{d}x + \eta_b \phi_i(b), \tag{3.40}$$

for $i = 2, 3, \ldots, N+1$.

We may combine Eqs. (3.38) and (3.40) into the linear system

$$A\mathbf{U} = \mathbf{b}, \tag{3.41}$$

where for $j = 1, 2, \ldots, N+1$,

$$A_{1,j} = \begin{cases} 1, & j = 1, \\ 0, & j \neq 1, \end{cases} \tag{3.42}$$

$$A_{i,j} = \int_a^b p \frac{\mathrm{d}\phi_i}{\mathrm{d}x} \frac{\mathrm{d}\phi_j}{\mathrm{d}x} + q\phi_i \frac{\mathrm{d}\phi_j}{\mathrm{d}x} + r\phi_i\phi_j \, \mathrm{d}x, \qquad i = 2, 3, \ldots, N+1, \tag{3.43}$$

and

$$b_1 = u_a, \tag{3.44}$$

$$b_i = \int_a^b f\phi_i \, \mathrm{d}x + \eta_b \phi_i(b), \qquad i = 2, 3, \ldots, N+1. \tag{3.45}$$

Appealing to Eq. (3.18), it is clear that

$$\phi_i(b) = \phi_i(x_{N+1}) = \begin{cases} 1, & i = N+1, \\ 0, & i \neq N+1, \end{cases}$$

and so this term only affects the equation generated with $i = N+1$. We may now write Eqs. (3.44) and (3.45), the entries of $\mathbf{b}$, as

$$b_1 = u_a, \tag{3.46}$$

$$b_i = \int_a^b f\phi_i \, \mathrm{d}x, \qquad i = 2, 3, \ldots, N, \tag{3.47}$$

$$b_{N+1} = \int_a^b f\phi_{N+1} \, \mathrm{d}x + \eta_b. \tag{3.48}$$

We now discuss how to assemble the algebraic equations given by Eq. (3.41), with a particular focus on handling the integration that is required.

3.4 Assembling the Algebraic Equations

We assemble the entries of the global matrix A and the global vector $\mathbf{b}$ that appear in the linear system Eq. (3.41) using the same strategy as in Sect. 2.8. We first evaluate those entries that are defined by integrals, namely Eqs. (3.43), (3.47) and (3.48), by decomposing these integrals into the sum of integrals over the elements in the mesh. Remembering that element k occupies $x_k \leq x \leq x_{k+1}$, we may write Eqs. (3.43), (3.47) and (3.48) as follows:

$$A_{i,j} = \sum_{k=1}^{N} \int_{x_k}^{x_{k+1}} p\frac{\mathrm{d}\phi_i}{\mathrm{d}x}\frac{\mathrm{d}\phi_j}{\mathrm{d}x} + q\phi_i\frac{\mathrm{d}\phi_j}{\mathrm{d}x} + r\phi_i\phi_j \,\mathrm{d}x,$$

$$i = 2, 3, \ldots, N+1, \qquad j = 1, 2, \ldots, N+1, \tag{3.49}$$

$$b_i = \sum_{k=1}^{N} \int_{x_k}^{x_{k+1}} f\phi_i \,\mathrm{d}x, \qquad i = 2, 3, \ldots, N, \tag{3.50}$$

$$b_{N+1} = \sum_{k=1}^{N} \int_{x_k}^{x_{k+1}} f\phi_{N+1} \,\mathrm{d}x + \eta_b. \tag{3.51}$$

We then complete the assembly of the linear system by setting the entries that are defined explicitly by Eqs. (3.42) and (3.46). We now explain this process in more detail.

3.4.1 The Contributions from Integrals

We see from Eq. (3.49) that the contribution to entry $A_{i,j}$ of the global matrix A, from integration over element k, is given by, for $i = 2, 3, \ldots, N+1$, $j = 1, 2, \ldots, N+1$:

$$\int_{x_k}^{x_{k+1}} p\frac{\mathrm{d}\phi_i}{\mathrm{d}x}\frac{\mathrm{d}\phi_j}{\mathrm{d}x} + q\phi_i\frac{\mathrm{d}\phi_j}{\mathrm{d}x} + r\phi_i\phi_j \,\mathrm{d}x.$$

In Sect. 2.8, we noted that the only basis functions that are nonzero on element k are $\phi_k(x)$ and $\phi_{k+1}(x)$. The integral above will therefore be zero unless both i and j take the values k or $k+1$. Consequently, the integral above generates nonzero contributions only to the entries $A_{k,k}$, $A_{k,k+1}$, $A_{k+1,k}$, $A_{k+1,k+1}$ of the global matrix A. These nonzero contributions may be stored in the local matrix $A_{\text{local}}^{(k)}$, of size 2×2, where entry $A_{\text{local},1,1}^{(k)}$ contributes to $A_{k,k}$; $A_{\text{local},1,2}^{(k)}$ contributes to $A_{k,k+1}$; $A_{\text{local},2,1}^{(k)}$ contributes to $A_{k+1,k}$; and $A_{\text{local},2,2}^{(k)}$ contributes to $A_{k+1,k+1}$. The entries of $A_{\text{local}}^{(k)}$ are given by:

$$A^{(k)}_{\text{local},\,1,1} = \int_{x_k}^{x_{k+1}} p\frac{\mathrm{d}\phi_k}{\mathrm{d}x}\frac{\mathrm{d}\phi_k}{\mathrm{d}x} + q\phi_k\frac{\mathrm{d}\phi_k}{\mathrm{d}x} + r\phi_k\phi_k \,\mathrm{d}x, \tag{3.52}$$

$$A^{(k)}_{\text{local},\,1,2} = \int_{x_k}^{x_{k+1}} p\frac{\mathrm{d}\phi_k}{\mathrm{d}x}\frac{\mathrm{d}\phi_{k+1}}{\mathrm{d}x} + q\phi_k\frac{\mathrm{d}\phi_{k+1}}{\mathrm{d}x} + r\phi_k\phi_{k+1} \,\mathrm{d}x, \tag{3.53}$$

$$A^{(k)}_{\text{local},\,2,1} = \int_{x_k}^{x_{k+1}} p\frac{\mathrm{d}\phi_{k+1}}{\mathrm{d}x}\frac{\mathrm{d}\phi_k}{\mathrm{d}x} + q\phi_{k+1}\frac{\mathrm{d}\phi_k}{\mathrm{d}x} + r\phi_{k+1}\phi_k \,\mathrm{d}x, \tag{3.54}$$

$$A^{(k)}_{\text{local},\,2,2} = \int_{x_k}^{x_{k+1}} p\frac{\mathrm{d}\phi_{k+1}}{\mathrm{d}x}\frac{\mathrm{d}\phi_{k+1}}{\mathrm{d}x} + q\phi_{k+1}\frac{\mathrm{d}\phi_{k+1}}{\mathrm{d}x} + r\phi_{k+1}\phi_{k+1} \,\mathrm{d}x. \tag{3.55}$$

Similarly, using Eqs. (3.50) and (3.51), the contribution to the entry b_i, $i = 2, 3, \ldots, N+1$, of the global vector $\mathbf{b}$ from integration over element k is given by

$$\int_{x_k}^{x_{k+1}} f\phi_i \,\mathrm{d}x.$$

Using the same argument as above, we see that this integral generates nonzero contributions only to the entries b_k and b_{k+1} of the global vector $\mathbf{b}$. We store these nonzero contributions in a local vector $\mathbf{b}^{(k)}_{\text{local}}$ with two entries. Entry $b^{(k)}_{\text{local},1}$ of this local vector contributes to entry b_k of the global vector $\mathbf{b}$, and entry $b^{(k)}_{\text{local},2}$ contributes to entry b_{k+1} of $\mathbf{b}$. The entries of $\mathbf{b}^{(k)}_{\text{local}}$ are given by:

$$b^{(k)}_{\text{local},\,1} = \int_{x_k}^{x_{k+1}} f\phi_k \,\mathrm{d}x, \tag{3.56}$$

$$b^{(k)}_{\text{local},\,2} = \int_{x_k}^{x_{k+1}} f\phi_{k+1} \,\mathrm{d}x. \tag{3.57}$$

We may therefore compute the integrals that appear in Eqs. (3.43), (3.47) and (3.48) by looping over all elements, calculating only the nonzero local contributions $A^{(k)}_{\text{local}}$ and $\mathbf{b}^{(k)}_{\text{local}}$ from each element, before using these local contributions to increment the appropriate entries of A and $\mathbf{b}$. We then complete the assembly of Eqs. (3.43), (3.47) and (3.48) by adding the extra algebraic term that appears in Eq. (3.48) to entry b_{N+1} of $\mathbf{b}$. However, before we may use this technique for assembling these entries of A and $\mathbf{b}$, we must first be able to evaluate the integrals that define the entries of $A^{(k)}_{\text{local}}$ and $\mathbf{b}^{(k)}_{\text{local}}$ on each element.

For some functions $p(x), q(x), r(x)$ and $f(x)$, the integrals given by Eqs. (3.52)–(3.57) may be evaluated analytically. This was the case for the simple example differential equation in Chap. 2, where $p(x) = 1, q(x) = 0, r(x) = 0$ and $f(x) = 2$. For more general functions, it may not be possible to evaluate these integrals analytically, and we must approximate these integrals numerically. This is one instance where the specification of the global basis functions through local basis functions on the canonical element $0 \le X \le 1$, described in Sect. 3.3.2.2, is useful, as we now describe.

We will see, in Sect. 3.5, that when making a numerical approximation to the integrals above, it is convenient to use the change of variable, given by Eq. (3.21), that maps between element k and the canonical element $0 \le X \le 1$. Writing

$$h = x_{k+1} - x_k, \qquad k = 1, 2, \ldots, N,$$

we may write the change of variable given by Eq. (3.21) as

$$x = x_k + hX, \qquad 0 \le X \le 1. \tag{3.58}$$

We now use this change of variable to write the entries of $A^{(k)}_{\text{local}}$ and $\mathbf{b}^{(k)}_{\text{local}}$, given by Eqs. (3.52)–(3.57), as integrals over the canonical element. On using Eq. (3.24) and (3.25), we may write ϕ_k and ϕ_{k+1} in terms of the local basis functions defined by Eqs. (3.22) and (3.23). Noting that

$$\frac{\mathrm{d}\phi_{\text{local},\,i}}{\mathrm{d}x} = \frac{1}{h}\frac{\mathrm{d}\phi_{\text{local},\,i}}{\mathrm{d}X}, \qquad i = 1, 2,$$

we may write the entries of $A^{(k)}_{\text{local}}$ and $\mathbf{b}^{(k)}_{\text{local}}$ as

$$\begin{aligned} A^{(k)}_{\text{local},\,i,j} &= h \int_0^1 \frac{1}{h^2} p(x_k + hX) \frac{\mathrm{d}\phi_{\text{local},\,i}}{\mathrm{d}X}\frac{\mathrm{d}\phi_{\text{local},\,j}}{\mathrm{d}X} + \\ &\quad \frac{1}{h} q(x_k + hX)\phi_{\text{local},\,i}(X)\frac{\mathrm{d}\phi_{\text{local},\,j}}{\mathrm{d}X} + \\ &\quad r(x_k + hX)\phi_{\text{local},\,i}(X)\phi_{\text{local},\,j}(X)\,\mathrm{d}X, \qquad i, j = 1, 2, \end{aligned} \tag{3.59}$$

$$\mathbf{b}^{(k)}_{\text{local},\,i} = h \int_0^1 f(x_k + hX)\phi_{\text{local},\,i}(X)\,\mathrm{d}X, \qquad i = 1, 2. \tag{3.60}$$

We will explain in Sect. 3.5 how these integrals may be approximated numerically.

3.4.2 *Completing the Assembly of the Linear System*

We noted in Sect. 2.8.2 that when looping over elements to calculate the contributions to A and $\mathbf{b}$ that arise from integrals, we introduce incorrect entries into the rows where the entries are given by Eqs. (3.42) and (3.46). We complete the assembly of the linear system by overwriting these incorrect entries and assigning the correct entries defined explicitly by Eqs. (3.42) and (3.46).

3.5 Approximating Integrals Using Numerical Quadrature

The integrals given by Eqs. (3.59) and (3.60) cannot be evaluated explicitly for general functions $p(x), q(x), r(x), f(x)$. Even if the integrals can be evaluated analytically, the integration may be tedious and time-consuming. Furthermore, if these functions were altered, we would then need to evaluate these integrals again using the new functions. Instead, we advise allocating this task to the computer, especially as a simple, flexible technique exists for approximating these integrals, namely *numerical quadrature*.

Suppose we wish to evaluate the integral

$$\int_0^1 g(X)\,\mathrm{d}X,$$

which is of the form given by Eqs. (3.59) and (3.60). When using numerical quadrature, the function $g(X)$ is evaluated at a collection of M points $X_1, X_2, \dots, X_M$ in the interval $0 \le X \le 1$, and the approximation to the integral is given by a weighted sum of these function evaluations:

$$\int_0^1 g(X)\,\mathrm{d}X \approx \sum_{m=1}^{M} w_m g(X_m), \tag{3.61}$$

where w_m, $m = 1, 2, \dots, M$, are known as the *quadrature weights* or simply the *weights*, and X_m, $m = 1, 2, \dots, M$, are known as the *quadrature points*.

Throughout this book, we shall assume that any integral that has been evaluated using quadrature has been approximated sufficiently accurately that it may be considered to be the true value of this integral. As such, in a slight abuse of mathematical notation, we shall use an equals sign, rather than the approximately equals sign that has been used in Eq. (3.61), when writing these integrals. Computational implementations will, by necessity, use an equals sign for an expression such as Eq. (3.61), and so the use of an equals sign in the text will aid clarity when presenting these implementations.

When designing a quadrature rule with M quadrature points, there are $2M$ quantities that can be chosen: M weights w_m and M quadrature points X_m. A polynomial of degree $2M-1$ is specified by the $2M$ coefficients of X^p, $p = 0, 1, 2, \dots, 2M-1$. Gaussian quadrature, one variety of quadrature, assigns the M weights and M quadrature points by insisting that Gaussian quadrature is exact for polynomials of degree $2M - 1$. It may be shown that for any value of M: (i) the Gaussian weights and quadrature points exist; (ii) the weights all take positive values; and (iii) the quadrature points all lie in the interval $0 < X < 1$. Weights and quadrature points for $M = 1, 2, 3, 4$ are given in Table 3.1. As would be expected, Gaussian quadrature becomes a more accurate approximation of the true value of the integral when more quadrature points are used.

Table 3.1 Gaussian quadrature points and weights for $M = 1, 2, 3, 4$ quadrature points

$M=1$		$M=2$		$M=3$		$M=4$	
X_m	w_m	X_m	w_m	X_m	w_m	X_m	w_m
$\frac{1}{2}$	1	$\frac{1}{2}\left(1-\frac{1}{\sqrt{3}}\right)$	$\frac{1}{2}$	$\frac{1}{2}\left(1-\sqrt{\frac{3}{5}}\right)$	$\frac{5}{18}$	$\frac{1}{2}\left(1-\sqrt{\frac{3}{7}+\frac{2}{7}\sqrt{\frac{6}{5}}}\right)$	$\frac{1}{72}\left(18-\sqrt{30}\right)$
		$\frac{1}{2}\left(1+\frac{1}{\sqrt{3}}\right)$	$\frac{1}{2}$	$\frac{1}{2}$	$\frac{4}{9}$	$\frac{1}{2}\left(1-\sqrt{\frac{3}{7}-\frac{2}{7}\sqrt{\frac{6}{5}}}\right)$	$\frac{1}{72}\left(18+\sqrt{30}\right)$
				$\frac{1}{2}\left(1+\sqrt{\frac{3}{5}}\right)$	$\frac{5}{18}$	$\frac{1}{2}\left(1+\sqrt{\frac{3}{7}-\frac{2}{7}\sqrt{\frac{6}{5}}}\right)$	$\frac{1}{72}\left(18+\sqrt{30}\right)$
						$\frac{1}{2}\left(1+\sqrt{\frac{3}{7}+\frac{2}{7}\sqrt{\frac{6}{5}}}\right)$	$\frac{1}{72}\left(18-\sqrt{30}\right)$

The quadrature approximation given by Eq. (3.61) may now be used to approximate the integrals given by Eqs. (3.59) and (3.60) as follows: for $i = 1, 2$, $j = 1, 2$:

$$\begin{aligned} A^{(k)}_{\text{local},\,i,j} = h \sum_{m=1}^{M} w_m \Bigg(& \frac{1}{h^2} p(x_k + hX_m) \left.\frac{\mathrm{d}\phi^{(k)}_{\text{local},i}}{\mathrm{d}X}\right|_{X=X_m} \left.\frac{\mathrm{d}\phi^{(k)}_{\text{local},j}}{\mathrm{d}X}\right|_{X=X_m} + \\ & \frac{1}{h} q(x_k + hX_M)\phi^{(k)}_{\text{local},i}(X_M) \left.\frac{\mathrm{d}\phi^{(k)}_{\text{local},j}}{\mathrm{d}X}\right|_{X=X_m} + \\ & r(x_k + hX_M)\phi^{(k)}_{\text{local},i}(X_M)\phi^{(k)}_{\text{local},j}(X_M)\Bigg), \end{aligned} \qquad (3.62)$$

and for $i = 1, 2$:

$$b^{(k)}_{\text{local},\,i} = h \sum_{m=1}^{M} w_m f(x_k + hX_M)\phi^{(k)}_{\text{local},i}(X_M), \qquad i = 1, 2. \qquad (3.63)$$

3.6 The Steps Required

In Chap. 2, we explained how to calculate the finite element solution of the simplest possible boundary value problem. A summary of the tasks required when calculating the finite element solution of this simple example was given in Sect. 2.9. In this chapter, we have developed the material presented in Chap. 2 to allow calculation of the finite element solution of a very general second-order boundary value problem. It is worth revisiting the summary of tasks given in Sect. 2.9 at this point. We see that, even for more general problems, this summary still forms a perfectly valid summary.

3.7 Computational Implementation

We demonstrate application of the material presented in this chapter using a model boundary value problem defined by

$$-\frac{\mathrm{d}}{\mathrm{d}x}\left(-x\frac{\mathrm{d}u}{\mathrm{d}x}\right)-(2x+3)\frac{\mathrm{d}u}{\mathrm{d}x}+(x+2)u=x^3\mathrm{e}^{2x}, \qquad 0.5<x<1, \quad (3.64)$$

together with a Dirichlet boundary condition at $x = 0.5$:

$$u=\mathrm{e}, \qquad x=0.5, \quad (3.65)$$

and a natural Neumann boundary condition at $x = 1$:

$$-x\frac{\mathrm{d}u}{\mathrm{d}x}=-5\mathrm{e}^2, \qquad x=1. \quad (3.66)$$

The boundary value problem above has solution

$$u(x)=\mathrm{e}^x\left(K_1+K_2x^3\right)+\mathrm{e}^{2x}\left(x^2-2x+2\right),$$

where

$$K_1=3\mathrm{e}-\frac{32}{31}\left(3\mathrm{e}+\frac{\sqrt{\mathrm{e}}}{4}\right),$$
$$K_2=\frac{8}{31}\left(3\mathrm{e}+\frac{\sqrt{\mathrm{e}}}{4}\right).$$

The boundary value problem above may be related to the general differential equation and boundary conditions given by Eq. (3.1)–(3.3) by setting

$$p(x)=-x,$$
$$q(x)=-(2x+3),$$
$$r(x)=x+2,$$
$$f(x)=x^3\mathrm{e}^{2x},$$
$$u_a=\mathrm{e}, \quad (3.67)$$
$$\eta_b=-5\mathrm{e}^2. \quad (3.68)$$

A MATLAB function for implementing the finite element method to solve the boundary value problem defined by Eqs. (3.64)–(3.66) is given in Listing 3.1. This function follows the outline for calculating a finite element solution given in Sect. 2.9, and variable names follow the mathematical notation used in Chaps. 2 and 3. One difference to the implementation given in Sect. 2.9 is the input required to the MATLAB function. In this chapter, we have not assumed that the nodes are equally spaced, and specifying only the number of elements N is no longer sufficient to define the mesh and basis functions. Instead, the vector of nodes used in the mesh is required as input. If N elements are used, this vector will contain $N + 1$ nodes. The vector of nodes supplied as input to the function is assumed to satisfy the conditions on the mesh set out in Sect. 3.3.1, that is (i) $x_1 = a$, (ii) $x_{N+1} = b$ and (iii) $x_{k+1} > x_k$ for $k = 1, 2, \ldots, N$. The function returns the vector $\mathbf{U}$ that specifies the finite element solution at each node.

We now give a commentary on how the MATLAB function in Listing 3.1 implements the material presented in this chapter. After taking a suitable vector of nodes, $\mathbf{x}$, we then define the functions $p(x)$, $q(x)$, $r(x)$, $f(x)$, the value of the Dirichlet boundary condition, u_a, and the value of the Neumann boundary condition, η_b, in lines 4–9 of the listing. The number of nodes in the mesh is equal to length of the vector $\mathbf{x}$. The number of elements, N, will be one less than the number of nodes: the value of N is set in line 12. After initialising all entries of A and $\mathbf{b}$ to zero in lines 15 and 16, we then loop over the elements in lines 20–26 calculating the local contributions to A and $\mathbf{b}$ from each element, before incrementing the global matrix A and global vector $\mathbf{b}$ appropriately. The local contributions on each element are calculated using the function `CalculateContributionOnElement` which is defined in lines 45–89 of the listing. In this function, we first deduce the element length, h, in line 49. We define the quadrature scheme we are using—that is, the number of quadrature points M, the weights and the quadrature points—in lines 52–54. We initialise the local contributions to $A_{\text{local}}^{(k)}$ and $\mathbf{b}_{\text{local}}^{(k)}$ to zero in lines 57–58 and then use this quadrature scheme to calculate the entries of $A_{\text{local}}^{(k)}$ and $\mathbf{b}_{\text{local}}^{(k)}$, given by Eqs. (3.62) and (3.63), by looping over the quadrature points in lines 61–89. Inside this loop, we first evaluate the local basis functions, $\phi_{\text{local},1}$, $\phi_{\text{local},2}$, the derivatives of the local basis functions and the functions $p(x)$, $q(x)$, $r(x)$, $f(x)$ at each quadrature point, before using these values to increment the entries of $A_{\text{local}}^{(k)}$ and $\mathbf{b}_{\text{local}}^{(k)}$ by weighting the contributions according to the quadrature rule.

Returning to the main body of code, the contribution from the Neumann boundary condition at $x = 1$ is added into b_{N+1} in line 30 using the value of η_b given by Eq. (3.68), before the Dirichlet boundary condition at $x = 0.5$ is imposed in lines 33–35 using the value of u_a given by Eq. (3.67). Finally, the linear system is solved in line 38, before the finite element solution is plotted in lines 39–41.

This MATLAB function may be called by typing, for example,

```
U = Chap3_CalculateFemForBvp(linspace(0.5, 1, 101));
```

into a MATLAB session to use a computational mesh with 101 nodes (and, therefore, 100 elements) with the nodes equally spaced between 0.5 and 1.

Listing 3.1 `Chap3_CalculateFemForBvp.m`, a MATLAB function for calculating the finite element solution of a boundary value problem of the form given in Sect. 3.1.

```
function U = Chap3_CalculateFemForBvp(x)

% define p(x), q(x), r(x), f(x), u_a, eta_b
p = @(x) -x;
q = @(x) -(2*x+3);
r = @(x) x+2;
f = @(x) x^3*exp(2*x);
u_a = exp(1);
eta_b = -5*exp(1)*exp(1);

% Deduce number of elements
N = length(x)-1;

% Initialise A and b to zero
A = zeros(N+1, N+1);
b = zeros(N+1, 1);

% Loop over elements calculating local contributions and
% inserting into linear system
for k = 1:N
    [A_local, b_local] = ...
        CalculateContributionOnElement(x, k, p, q, r, f);

    A(k:k+1, k:k+1) = A(k:k+1, k:k+1) + A_local;
    b(k:k+1) = b(k:k+1) + b_local;
end

% Add contributions from Neumann boundary conditions into
% linear system
b(N+1) = b(N+1) + eta_b;

% Set Dirichlet boundary conditions in row 1
A(1,:) = 0;
A(1,1) = 1;
b(1) = u_a;

% Solve linear system and plot FE solution
U = A\b;
plot(x,U,'-o')
xlabel('x')
ylabel('U')



function [A_local, b_local] = ...
    CalculateContributionOnElement(x, k, p, q, r, f)

% Calculate length of element
h = x(k+1) - x(k);

% Define quadrature scheme
M = 3;
QuadWeights = [5/18, 4/9, 5/18];
QuadPoints = [0.5*(1-sqrt(0.6)), 0.5, 0.5*(1+sqrt(0.6))];

```

```
% Initialise local contributions to A and b to zero
A_local = zeros(2,2);
b_local = zeros(2,1);

% Loop over quadrature points
for m = 1:M
    X = QuadPoints(m);

    % Evaluate local basis functions and their derivatives
    phi_local = [1-X, X];
    phi_prime_local = [-1, 1];

    % Evaluate p, q, r, f at quadrature points
    p_val = p(x(k)+h*X);
    q_val = q(x(k)+h*X);
    r_val = r(x(k)+h*X);
    f_val = f(x(k)+h*X);

    % Increment entries of A_local and b_local
    % with suitably weighted function evaluations
    for i=1:2
        for j=1:2
            A_local(i,j) = A_local(i,j) + ...
                h*QuadWeights(m)*( ...
                p_val/h/h*phi_prime_local(i)* ...
                phi_prime_local(j) + ...
                q_val/h*phi_local(i)* ...
                phi_prime_local(j) + ...
                r_val*phi_local(i)*phi_local(j));
        end
        b_local(i) = b_local(i) + h*QuadWeights(m)* ...
            f_val*phi_local(i);
    end
end
```

3.8 Robin Boundary Conditions

In Sect. 3.1, we defined a general second-order differential equation, Eq. (3.1), that was subject to one Dirichlet boundary condition, Eq. (3.2), and one natural Neumann boundary condition, Eq. (3.3). Many boundary value problems can be written in the form of Eq. (3.1), subject to some combination of Dirichlet and Neumann boundary conditions, and so this can be thought of as a reasonably general boundary value problem. There is, however, a third variety of boundary condition—known as a *Robin* boundary condition—that relates both the function value *and* the derivative of the function on the boundary.

We demonstrate how Robin boundary conditions may be handled by the finite element method using a slight modification of the boundary value problem given in Sect. 3.1. We use the differential equation given by Eq. (3.1) on the interval $a < x < b$ and the Dirichlet boundary condition at $x = a$ given by Eq. (3.2). Rather than using the Neumann boundary condition given by Eq. (3.3), we specify a Robin boundary condition at $x = b$. The differential equation and boundary conditions are then given by

$$-\frac{\mathrm{d}}{\mathrm{d}x}\left(p(x)\frac{\mathrm{d}u}{\mathrm{d}x}\right) + q(x)\frac{\mathrm{d}u}{\mathrm{d}x} + r(x)u = f(x), \tag{3.69}$$

$$u = u_a, \qquad x = a, \tag{3.70}$$

$$p(x)\frac{\mathrm{d}u}{\mathrm{d}x} + \alpha u = \eta_b, \qquad x = b, \tag{3.71}$$

where $p(x), q(x), r(x), f(x)$ are given functions, and u_a, η_b, α are given constants.

3.8.1 The Weak Formulation

As before, the first step when computing the finite element solution of Eqs. (3.69)–(3.71) is to write the problem in weak form. We have a Dirichlet boundary condition at $x = a$, and so the space $H_0^1(a, b)$ for this boundary value problem comprises those functions $v(x) \in H^1(a, b)$ that also satisfy $v(a) = 0$. We then proceed as in Sect. 3.2.2. We multiply the given differential equation by a test function $v(x) \in H_0^1(a, b)$ and integrate over the interval $a < x < b$. After integrating by parts, we obtain

$$\left[-p\frac{\mathrm{d}u}{\mathrm{d}x}v\right]_a^b + \int_a^b p\frac{\mathrm{d}u}{\mathrm{d}x}\frac{\mathrm{d}v}{\mathrm{d}x} + q\frac{\mathrm{d}u}{\mathrm{d}x}v + ruv \, \mathrm{d}x = \int_a^b fv \, \mathrm{d}x.$$

Using $v(a) = 0$ (a consequence of $v(x) \in H_0^1(a, b)$) and the Robin boundary condition, Eq. (3.71), the equation above may be written as

$$\int_a^b p\frac{\mathrm{d}u}{\mathrm{d}x}\frac{\mathrm{d}v}{\mathrm{d}x} + q\frac{\mathrm{d}u}{\mathrm{d}x}v + ruv \, \mathrm{d}x + \alpha u(b)v(b) = \int_a^b fv \, \mathrm{d}x + \eta_b v(b). \tag{3.72}$$

We then define the weak solution of Eqs. (3.69)–(3.71) by:

find $u(x) \in H_E^1(a, b)$ such that Eq. (3.72) holds for all $v(x) \in H_0^1(a, b)$,

where as defined by Eq. (3.10), $H_E^1(a, b)$ is the set of $u(x) \in H^1(a, b)$ that also satisfy the Dirichlet boundary condition given by Eq. (3.70).

3.8.2 *The Finite Element Solution*

Having derived the weak solution, we then specify the set S^h, and the subsets of this set S_E^h and S_0^h, as in Sect. 3.3.2. We then define the finite element solution, $U(x)$, of Eqs. (3.69)–(3.71) by

find $U(x) \in S_E^h$ such that

$$\int_a^b p\frac{\mathrm{d}U}{\mathrm{d}x}\frac{\mathrm{d}\phi_i}{\mathrm{d}x} + q\frac{\mathrm{d}U}{\mathrm{d}x}\phi_i + rU\phi_i \;\mathrm{d}x + \alpha U(b)\phi_i(b) = \int_a^b f\phi_i \;\mathrm{d}x + \eta_b\phi_i(b), \tag{3.73}$$

for all $\phi_i(x) \in S_0^h$.

The only difference between Eq. (3.73) and the finite element solution given by Eq. (3.35) is the extra term $\alpha U(b)\phi_i(b)$ on the left-hand side of Eq. (3.73). This extra term is straightforward to handle: using Eq. (3.18), we see that

$$\phi_i(b) = \begin{cases} 1, & i = N+1, \\ 0, & i \neq N+1, \end{cases}$$

and so this term will contribute only to row $N+1$ of the linear system. Furthermore, using Eq. (3.20),

$$U(b) = U(x_{N+1}) = U_{N+1}.$$

The extra term in Eq. (3.73) therefore adds an extra term αU_{N+1} to the left-hand side of equation generated by using $\phi_i = \phi_{N+1}$ in Eq. (3.73), that is row $N+1$ of the linear system given by Eq. (3.41). This may be implemented very easily by adding the quantity α to the value of $A_{N+1,N+1}$ defined earlier by Eq. (3.43).

3.9 A Bound on the Accuracy of the Solution

In this chapter, we have set out the theory necessary to apply the finite element method to a general linear, second-order, boundary value problem. We have not yet, however, said anything about whether the finite element solution is a good approximation to the true solution, or how this approximation improves as the number of elements in the mesh is increased. We now provide—without proof—a theorem on the accuracy of the finite element solution.

Theorem 3.1 *Let $p(x), q(x), r(x), f(x)$ be functions such that $p(x), q(x), r(x), f(x)$ and $\frac{\mathrm{d}p}{\mathrm{d}x}$ are continuous functions on the interval $a < x < b$. Suppose the second-order differential equation, defined on $a < x < b$, and given by*

$$-\frac{\mathrm{d}}{\mathrm{d}x}\left(p(x)\frac{\mathrm{d}u}{\mathrm{d}x}\right)+q(x)\frac{\mathrm{d}u}{\mathrm{d}x}+r(x)u=f(x),$$

has a unique solution, subject to suitable given Dirichlet, Neumann or Robin boundary conditions. Let $U(x)$ be the finite element solution, calculated on a uniform mesh of elements of length h, that is a linear approximation to the solution on each element. Define the error, $E(x)$, by

$$E(x)=U(x)-u(x),$$

and the L^2 norm of the error by

$$\|E\|_{L^2}=\sqrt{\int_a^b (E(x))^2\,\mathrm{d}x}.$$

There then exists a constant $C>0$, independent of the mesh used, such that, as the element size $h\to 0$, the error satisfies

$$\|E\|_{L^2}=Ch^2.$$

The constant C that appears in Theorem 3.1 is not easy to obtain. If C is not known, the reader may ask, quite reasonably, what the purpose of this error bound is. There is a very compelling answer to this question. Suppose we calculate the finite element solution on a mesh of elements with constant length h_1. Suppose we then compute a second finite element solution on a mesh where each element of the original mesh is divided into two, so that the length of each element in the new mesh is $h_2=h_1/2$. As the error in the L^2 norm depends on h^2, we see that halving the element length will reduce the L^2 norm of the error to a quarter of its original value. As such, we can use the error bound to assess the relative error between these two finite element solutions.

The error bound in Theorem 3.1 may also be used to confirm the efficiency of the implementation of the finite element method when it is used to solve a model problem with known solution, as was first introduced in Exercise 2.2 at the end of Chap. 2. Taking the logarithm of both sides of the error bound in Theorem 3.1 gives

$$\log\|E\|_{L^2}=\log C+2\log h. \tag{3.74}$$

Noting that $\|E\|_{L^2}$ may be calculated using

$$\|E\|_{L^2}=\sqrt{\sum_{k=1}^{N}\int_{x_k}^{x_{k+1}}(U(x)-u(x))^2\,\mathrm{d}x}, \tag{3.75}$$

we may easily compute $\|E\|_{L^2}$ for a given h, using quadrature if the integration is difficult, or impossible, to evaluate analytically. See Exercise 2.2 at the end of Chap. 2 for more details on performing this computation. Plotting $\log \|E\|_{L^2}$ against $\log h$ for a range of values of h, we see, from Eq. (3.74), that this graph should have slope 2 as $h \to 0$. If this slope is less than 2, then the implementation is not optimal.

3.10 Exercises

3.1 In this exercise, we will develop code that may be used to calculate the finite element solution of the boundary value problem, given by the differential equation

$$-\frac{\mathrm{d}}{\mathrm{d}x}\left(-4\frac{\mathrm{d}u}{\mathrm{d}x}\right) + 8\frac{\mathrm{d}u}{\mathrm{d}x} + 5u = \beta\left(\sin x + 8\cos x\right), \qquad -\pi < x < \pi,$$

subject to one natural Neumann boundary condition

$$-4\frac{\mathrm{d}u}{\mathrm{d}x} = -4\alpha, \qquad x = -\pi,$$

and one Dirichlet boundary condition

$$u = 2, \qquad x = \pi,$$

where α and β are constants. This boundary value problem has solution

$$u(x) = 2\mathrm{e}^{-x}\left(\mathrm{e}^{\pi}\sin\frac{x}{2} + \frac{(\alpha + 1 - 2\mathrm{e}^{2\pi})}{\mathrm{e}^{\pi}}\cos\frac{x}{2}\right) + \beta\sin x.$$

You may want to develop your code for calculating the finite element solution in the subparts below based on the overview of the steps required given in Sect. 2.9, and Listing 3.1.

The integrals required to assemble the global linear system for this example problem may all be evaluated analytically without too much difficulty. The reader may decide for themselves whether to evaluate these integrals analytically, or to approximate them numerically using Gaussian quadrature.

(a) Write down the weak formulation of the boundary value problem. Be sure to state clearly how the Dirichlet boundary conditions are handled, and what conditions are required on the test functions.
(b) Define the finite element solution and the linear system that determines the coefficients of the basis functions in the finite element solution.
(c) Write code to calculate the finite element solution of the boundary value problem in the case where $\alpha = 0$ and $\beta = 0$. Note that some terms appearing in the linear system do not need to be computed for these values of α and β.

(d) Extend the code you wrote in part (c) so that it may be used to calculate the finite element solution of the boundary value problem in the case where $\alpha = 5$ and $\beta = 0$.

(e) Extend the code you wrote in part (d) so that it may be used to calculate the finite element solution of the boundary value problem in the case where $\alpha = 5$ and $\beta = 1$.

(f) Investigate the use of a nonuniform mesh with the code that you wrote for part (e).

In parts (c)–(f) above be sure to validate your finite element solution by comparison with the true solution.

3.2 In this exercise, we will work with the differential equation

$$-\frac{\mathrm{d}}{\mathrm{d}x}\left((1+x)^2\frac{\mathrm{d}u}{\mathrm{d}x}\right) + \left(2+x-x^2\right)\frac{\mathrm{d}u}{\mathrm{d}x} + \frac{2x-1}{4}u = \left((1+x)^2 + \frac{2x-1}{4}\right)\sin x - x(1+x)\cos x, \quad 0 < x < \pi.$$

This differential equation has general solution

$$u(x) = \sqrt{1+x}\left(K_1 + K_2\mathrm{e}^{-x}\right) + \sin x,$$

for constants K_1 and K_2 that are determined from the boundary conditions. We will now consider different selections of boundary conditions and derive the finite element solution in each case. Remember that the weak formulation will change when the boundary conditions are changed.

We suggest that the reader uses Gaussian quadrature to approximate the integrals required to assemble the global linear system in this exercise.

(a) Compute the finite element approximation to the differential equation when $u(x)$ satisfies the Dirichlet boundary conditions

$$u(0) = 2, \qquad u(\pi) = \sqrt{1+\pi}.$$

In this case, the constants K_1, K_2 that appear in the general solution are given by

$$K_1 = \frac{1-2\mathrm{e}^{-\pi}}{1-\mathrm{e}^{-\pi}}, \qquad K_2 = \frac{1}{1-\mathrm{e}^{-\pi}}.$$

Check that your finite element solution is correct by comparison with the true solution.

Calculate the finite element solution on several uniform meshes, and calculate the L^2 norm of the error for each mesh. Confirm the result of Theorem 3.1, namely that the graph of $\log \|E\|_{L^2}$ against $\log h$, where h is the element size, has slope 2 as $h \to 0$.

(b) Compute the finite element approximation to the differential equation when $u(x)$ satisfies one Dirichlet boundary condition:

$$u = 0, \qquad x = 0,$$

and one natural Neumann boundary condition

$$(1+x)^2 \frac{\mathrm{d}u}{\mathrm{d}x} = 0, \qquad x = \pi.$$

In this case, the constants K_1, K_2 that appear in the general solution are given by

$$K_1 = \frac{2\sqrt{1+\pi}}{1+\mathrm{e}^{-\pi}(1+2\pi)}, \qquad K_2 = -\frac{2\sqrt{1+\pi}}{1+\mathrm{e}^{-\pi}(1+2\pi)}.$$

Check that your finite element solution is correct by comparison with the true solution.

Calculate the finite element solution on several uniform meshes, and calculate the L^2 norm of the error for each mesh. Confirm the result of Theorem 3.1, as in part (a).

3.3 Calculate the finite element solution of the partial differential equation

$$\frac{\mathrm{d}^2u}{\mathrm{d}x^2} - u = 0, \qquad 0 < x < 1,$$

subject to one Robin boundary condition:

$$\frac{\mathrm{d}u}{\mathrm{d}x} + u = 2, \qquad x = 0,$$

and one Dirichlet boundary condition:

$$u = \mathrm{e}, \qquad x = 1.$$

Confirm your finite element solution is correct by comparison with the true solution, $u = \mathrm{e}^x$.

3.4 A partial differential equation is defined, for $x^2 + y^2 < 1$, by

$$-\left(\frac{\partial^2 u}{\partial x^2} + \frac{\partial^2 u}{\partial x^2}\right) = 1,$$

subject to the Dirichlet boundary conditions

$$u(x, y) = 0, \qquad x^2 + y^2 = 1.$$

This partial differential equation may be written in, cylindrical polar coordinates, as the ordinary differential equation

$$-\frac{1}{r}\frac{\mathrm{d}}{\mathrm{d}r}\left(r\frac{\mathrm{d}u}{\mathrm{d}r}\right) = 1, \tag{3.76}$$

subject to the boundary conditions

$$\frac{\mathrm{d}u}{\mathrm{d}r} = 0, \qquad r = 0, \tag{3.77}$$

$$u = 0, \qquad r = 1, \tag{3.78}$$

where $r^2 = x^2 + y^2$. This ordinary differential equation has solution $u(r) = (1 - r^2)/4$, and so the solution of the partial differential equation is

$$u(x, y) = \frac{1}{4}\left(1 - x^2 - y^2\right).$$

By writing Eq. (3.76) in the form of Eq. (3.1), calculate the finite element solution of Eqs. (3.76)–(3.78). Verify that your solution is correct by comparison with the true solution.

3.5 There is no guarantee that a solution exists to a general boundary value problem. Furthermore, if a solution does exist, then it may not be unique. Should we attempt to compute the finite element solution to a boundary value problem without a unique solution, we would anticipate problems. In this exercise, we investigate this concept using a simple example. The differential equation

$$\frac{\mathrm{d}^2u}{\mathrm{d}x^2} + u = 0, \qquad 0 < x < \pi,$$

has general solution $u(x) = K_1 \sin x + K_2 \cos x$, where K_1, K_2 are constants determined from the boundary conditions.

(a) When $u(x)$ satisfies the Dirichlet boundary conditions

$$u(0) = u(\pi) = 0,$$

the differential equation has the nonunique solution $u(x) = K_1 \sin x$ for any constant K_1. Attempt to calculate the finite element solution to this boundary value problem. How does the solution vary with the number of elements N?

(b) When $u(x)$ satisfies the Dirichlet boundary conditions

$$u(0) = 1, \qquad u(\pi) = 0,$$

the differential equation has no solution. Attempt to calculate the finite element solution to this boundary value problem. How does the solution vary with the number of elements N?

Chapter 4
Higher Order Basis Functions

We claimed, in Chap. 1, that the finite element method is a very flexible technique for calculating the numerical solution of differential equations. One justification for this claim is that the finite element method allows the solution of a differential equation to be approximated, on each element of the computational mesh, using a polynomial of any degree chosen by the user. The material presented in earlier chapters of this book has focused entirely on finite element solutions that are linear approximations on each element. In this chapter, we extend this material to allow a general polynomial approximation on each element.

The key to defining higher order global basis functions is to provide a mapping between each element of the mesh and a canonical element, and to specify local basis functions of the required degree on this canonical element. This approach was introduced for linear basis functions in Chap. 3, and in this chapter, we will show how it may be extended to higher order basis functions.

4.1 A Model Differential Equation

We illustrate the use of a higher degree of approximating polynomial on each element using the same boundary value problem that was used in Chap. 3. That is, for $a < x < b$,

$$-\frac{\mathrm{d}}{\mathrm{d}x}\left(p(x)\frac{\mathrm{d}u}{\mathrm{d}x}\right) + q(x)\frac{\mathrm{d}u}{\mathrm{d}x} + r(x)u = f(x), \tag{4.1}$$

for given functions $p(x)$, $q(x)$, $r(x)$ and $f(x)$, where $p(x) \neq 0$ for $a < x < b$, subject to a Dirichlet boundary condition at $x = a$:

$$u = u_a, \qquad x = a, \tag{4.2}$$

J. Whiteley, *Finite Element Methods*, Mathematical Engineering,
DOI 10.1007/978-3-319-49971-0_4

for some given constant u_a, and a natural Neumann boundary condition at $x = b$:

$$p(x)\frac{\mathrm{d}u}{\mathrm{d}x} = \eta_b, \qquad x = b, \tag{4.3}$$

for a given constant η_b.

The first step when calculating a finite element solution of a differential equation is to derive the weak formulation of the differential equation. This step is independent of the basis functions used and will therefore be identical to that derived in Sect. 3.2.2, re-stated below:

find $u(x) \in H_E^1(a, b)$ such that

$$\int_a^b p\frac{\mathrm{d}u}{\mathrm{d}x}\frac{\mathrm{d}v}{\mathrm{d}x} + q\frac{\mathrm{d}u}{\mathrm{d}x}v + ruv \,\mathrm{d}x = \int_a^b fv \,\mathrm{d}x + \eta_b v(b), \tag{4.4}$$

for all $v(x) \in H_0^1(a, b)$.

4.2 Quadratic Basis Functions

We now describe how to calculate the finite element solution of the model boundary value problem, using a quadratic approximation to the solution on each element.

4.2.1 *The Mesh*

As in Chaps. 2 and 3, we partition the domain on which the differential equation is defined, i.e. the interval $a < x < b$, into N elements. We do not make any assumptions on the size of the elements, allowing a nonuniform mesh to be used.

In earlier chapters, when calculating a finite element solution that was a linear approximation to the solution on each element, we saw that a mesh comprising N elements contained $N + 1$ nodes, with all nodes located on the boundaries of the elements. When using a quadratic approximation to the solution on each element, we will require, in addition to the nodes on the boundary of each element, an additional node inside each element—a total of N extra nodes. The nodes inside each element are often referred to as *interior nodes*. Our mesh now contains $2N + 1$ nodes, $x_1, x_2, \ldots, x_{2N+1}$. As before, we set the first and last node to lie on the boundary, so that $x_1 = a$, and $x_{2N+1} = b$, and insist that $x_{j+1} > x_j$ for $j = 1, 2, \ldots, 2N$.

Using the conditions on the nodes set out above, element k occupies the interval $x_{2k-1} \leq x \leq x_{2k+1}$ for $k = 1, 2, \ldots, N$. We introduce one further condition on the nodes: the node x_{2k} is placed at the centre of element k for $k = 1, 2, \ldots, N$, so that

$$x_{2k} = 0.5(x_{2k-1} + x_{2k+1}). \tag{4.5}$$

The nodes and elements for a suitable computational mesh are illustrated in Fig. 4.1.

4.2.2 The Definition of Quadratic Basis Functions

We now define quadratic basis functions on the mesh illustrated in Fig. 4.1. We follow the same approach as in Sect. 3.3.2.2 and specify global basis functions through local basis functions defined on the canonical element $0 \leq X \leq 1$.

As element k occupies the interval $x_{2k-1} \leq x \leq x_{2k+1}$, a suitable change of variable that maps between element k and the canonical element is given by

$$x = x_{2k-1} + (x_{2k+1} - x_{2k-1})X, \qquad 0 \leq X \leq 1. \tag{4.6}$$

Local basis functions on the canonical element are defined by

$$\phi_{\text{local},1}(X) = 2\left(\frac{1}{2} - X\right)(1 - X), \tag{4.7}$$

$$\phi_{\text{local},2}(X) = 4X(1 - X), \tag{4.8}$$

$$\phi_{\text{local},3}(X) = 2X\left(X - \frac{1}{2}\right). \tag{4.9}$$

These local basis functions are illustrated in Fig. 4.2. We now define the global basis functions on element k by:

$$\phi_{2k-1}(X) = \phi_{\text{local},1}(X), \qquad 0 \leq X \leq 1, \tag{4.10}$$

$$\phi_{2k}(X) = \phi_{\text{local},2}(X), \qquad 0 \leq X \leq 1, \tag{4.11}$$

$$\phi_{2k+1}(X) = \phi_{\text{local},3}(X), \qquad 0 \leq X \leq 1, \tag{4.12}$$

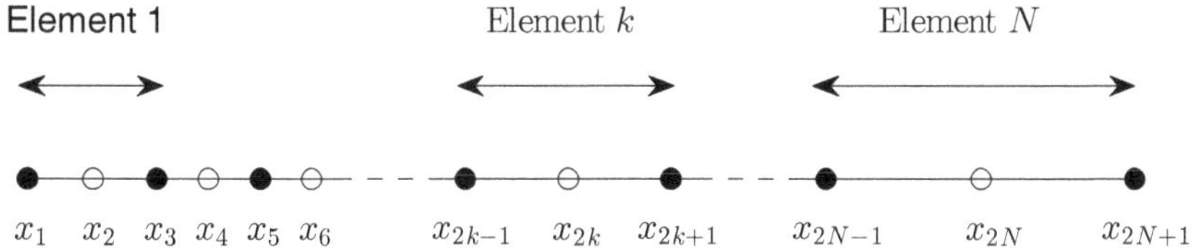

Fig. 4.1 The nodes and elements for a mesh that may be used to calculate a finite element solution that is a quadratic approximation to the true solution on each element. Nodes on the boundary of an element are denoted by a *filled circle*, and nodes in the interior of an element are denoted by an *open circle*

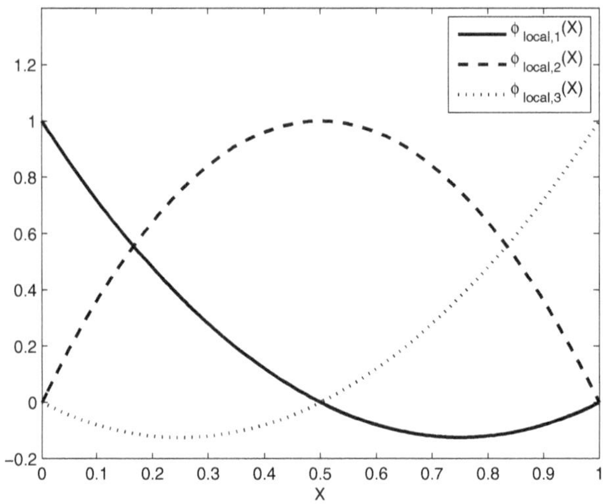

Fig. 4.2 The local basis functions $\phi_{\text{local},1}(X)$, $\phi_{\text{local},2}(X)$ and $\phi_{\text{local},3}(X)$ defined by Eqs. (4.7)–(4.9)

$$\phi_j(X) = 0, \qquad 0 \le X \le 1, \qquad j \ne 2k-1, j \ne 2k, j \ne 2k+1. \tag{4.13}$$

The definition of global basis functions by Eqs. (4.6)–(4.13) may be used on any element $k = 1, 2, \ldots, N$, in the mesh, fully specifying each of the global basis functions $\phi_1(x), \phi_2(x), \ldots, \phi_{2N+1}(x)$. We now describe some of the features of these global basis functions. Remember, from Eq. (4.5), that the interior node of each element is placed at the midpoint of that element. Using the change of variable given by Eq. (4.6), the local basis functions given by Eqs. (4.7)–(4.9) may be written in terms of the original variable x. On element k, we may then write the global basis functions given by Eqs. (4.10)–(4.13) as, for $x_{2k-1} \le x \le x_{2k+1}$:

$$\phi_{2k-1}(x) = \frac{(x_{2k} - x)(x_{2k+1} - x)}{(x_{2k} - x_{2k-1})(x_{2k+1} - x_{2k-1})}, \tag{4.14}$$

$$\phi_{2k}(x) = \frac{(x - x_{2k-1})(x_{2k+1} - x)}{(x_{2k} - x_{2k-1})(x_{2k+1} - x_{2k})}, \tag{4.15}$$

$$\phi_{2k+1}(x) = \frac{(x - x_{2k-1})(x - x_{2k})}{(x_{2k+1} - x_{2k-1})(x_{2k+1} - x_{2k})}, \tag{4.16}$$

$$\phi_j(x) = 0, \qquad j \ne 2k-1, j \ne 2k, j \ne 2k+1. \tag{4.17}$$

The basis functions $\phi_{2k-1}(x), \phi_{2k}(x), \phi_{2k+1}(x)$ are plotted in Fig. 4.3. We note that the global basis functions fall into two categories. The first category contains basis functions such as $\phi_{2k-1}(x)$ and $\phi_{2k+1}(x)$ that are nonzero in two neighbouring

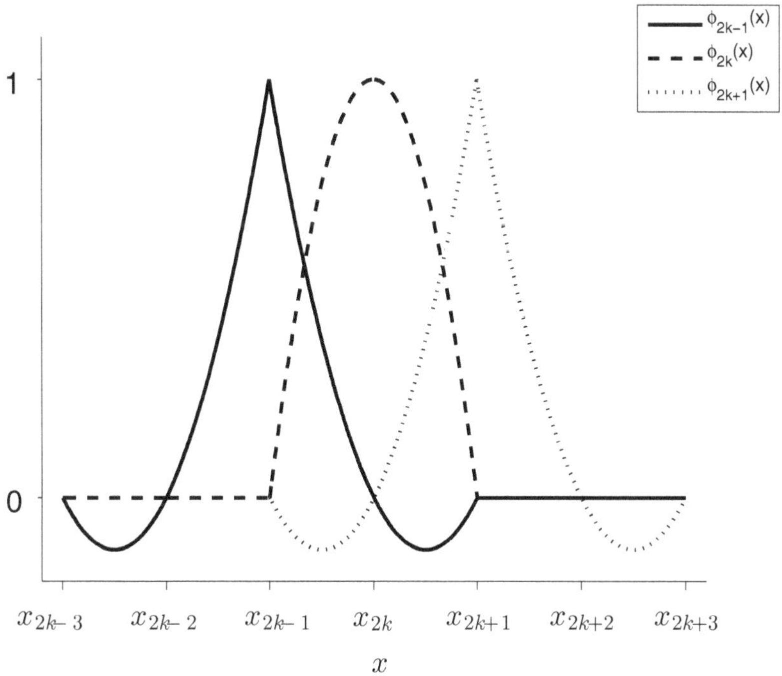

Fig. 4.3 The basis functions $\phi_{2k-1}(x)$, $\phi_{2k}(x)$ and $\phi_{2k+1}(x)$ defined by Eqs. (4.6)–(4.13)

elements. The second category is the basis functions such as $\phi_{2k}(x)$ that are nonzero in only one element.

Using Eqs. (4.14)–(4.17) we see that, on element k,

$$\phi_{2k-1}(x_{2k-1}) = 1, \qquad \phi_{2k-1}(x_{2k}) = 0, \qquad \phi_{2k-1}(x_{2k+1}) = 0, \tag{4.18}$$

$$\phi_{2k}(x_{2k-1}) = 0, \qquad \phi_{2k}(x_{2k}) = 1, \qquad \phi_{2k}(x_{2k+1}) = 0, \tag{4.19}$$

$$\phi_{2k+1}(x_{2k-1}) = 0, \qquad \phi_{2k+1}(x_{2k}) = 0, \qquad \phi_{2k+1}(x_{2k+1}) = 1, \tag{4.20}$$

and, if $j \neq 2k-1$, $j \neq 2k$ and $j \neq 2k+1$,

$$\phi_j(x_{2k-1}) = 0, \qquad \phi_j(x_{2k}) = 0, \qquad \phi_j(x_{2k+1}) = 0. \tag{4.21}$$

As this is true for all elements, we may deduce that, for all basis functions $\phi_j(x)$, $j = 1, 2, \ldots, 2N+1$, we have:

$$\phi_j(x_i) = \begin{cases} 1, & i = j, \\ 0, & i \neq j, \end{cases} \tag{4.22}$$

and so the quadratic global basis function $\phi_j(x)$ takes the value 1 at the node where $x = x_j$, and the value zero at all other nodes.

4.2.3 The Finite Element Solution

In Chap. 3, we defined the set S^h through the linear basis functions specified in Sect. 3.3.2. We now modify the definition of this set to take account of the quadratic basis functions specified in Sect. 4.2.2. The set S^h is now defined to be those functions $V(x)$ that can be written in the form

$$V(x) = \sum_{j=1}^{2N+1} V_j \phi_j(x), \tag{4.23}$$

where $\phi_j(x)$, $j = 1, 2, \ldots, 2N+1$, are the quadratic basis functions that are given by Eqs. (4.6)–(4.13). The subsets S_E^h and S_0^h of the set S^h are then defined in a similar manner to Sect. 3.3.3: the set S_E^h contains functions of the form given by Eq. (4.23) that satisfy all Dirichlet boundary conditions, and the set S_0^h contains functions of the form given by Eq. (4.23) that take the value zero at all points where Dirichlet boundary conditions are applied.

Having defined the sets S_E^h and S_0^h, we may now define the finite element solution, $U(x)$, by the following modification of the weak solution given by Eq. (4.4):

find $U(x) \in S_E^h$ such that

$$\int_a^b p \frac{\mathrm{d}U}{\mathrm{d}x}\frac{\mathrm{d}\phi_i}{\mathrm{d}x} + q\frac{\mathrm{d}U}{\mathrm{d}x}\phi_i + rU\phi_i \,\mathrm{d}x = \int_a^b f\phi_i \,\mathrm{d}x + \eta_b \phi_i(b), \tag{4.24}$$

for all $\phi_i(x) \in S_0^h$.

4.2.4 The System of Algebraic Equations

We write our finite element solution as a linear sum of the basis functions defined by Eqs. (4.6)–(4.13):

$$U(x) = \sum_{j=1}^{2N+1} U_j \phi_j(x), \tag{4.25}$$

and note that $U(x)$, as written above, lies in the set S^h defined in Sect. 4.2.3. Using Eq. (4.22), we have

$$U(x_i) = U_i, \qquad i = 1, 2, \ldots, 2N+1, \tag{4.26}$$

and so, as in earlier chapters, U_i, for $i = 1, 2, \ldots, 2N+1$, is the finite element solution at the node where $x = x_i$.

The finite element solution given by Eq. (4.25) contains $2N + 1$ unknown quantities $U_1, U_2, \ldots, U_{2N+1}$, and so we will need $2N + 1$ equations to determine these values. As in Chaps. 2 and 3, these equations will arise from two sources: (i) demanding that the finite element solution satisfies the Dirichlet boundary conditions and (ii) substituting suitable basis functions $\phi_i(x) \in S_0^h$ into the integral given by Eq. (4.24). We now explain how both categories of equations may be generated.

We first derive an algebraic equation that ensures that the finite element solution given by Eq. (4.25) satisfies all Dirichlet boundary conditions, thus satisfying the condition that $U(x)$ lies in the set S_E^h. The model problem contains one Dirichlet boundary condition, given by Eq. (4.2), which applies at the boundary $x = a$. This boundary condition therefore applies at the node where $x = x_1$. Using Eq. (4.26), we see that the finite element solution satisfies the Dirichlet boundary condition at $x = x_1$ if

$$U_1 = u_a, \tag{4.27}$$

which is our first algebraic equation.

We now explain how a further $2N$ algebraic equations may be generated by substituting suitable basis functions $\phi_i(x) \in S_0^h$ into the integral condition given by Eq. (4.24). Noting that the Dirichlet boundary condition given by Eq. (4.2) is applied at the node where $x = x_1$, and observing that $\phi_i(x_1) = 0$ for $i = 2, 3, \ldots, 2N+1$, we see that these $2N$ basis functions take the value zero where Dirichlet boundary conditions are applied. These functions are therefore members of the set S_0^h and suitable candidates for use in Eq. (4.24). Substituting these basis functions into Eq. (4.24) yields, after some manipulation, a total of $2N$ algebraic equations:

$$\sum_{j=1}^{2N+1} \left(\int_a^b p \frac{\mathrm{d}\phi_j}{\mathrm{d}x} \frac{\mathrm{d}\phi_i}{\mathrm{d}x} + q \frac{\mathrm{d}\phi_j}{\mathrm{d}x} \phi_i + r \phi_j \phi_i \, \mathrm{d}x \right) U_j = \int_a^b f \phi_i \, \mathrm{d}x + \eta_b \phi_i(b), \tag{4.28}$$

for $i = 2, 3, \ldots, 2N+1$.

When defining the mesh in Sect. 4.2.1, we set $x_{2N+1} = b$. Using Eq. (4.22), we may write

$$\begin{aligned} \phi_i(b) &= \phi_i(x_{2N+1}) \\ &= \begin{cases} 1, & i = 2N+1, \\ 0, & i \neq 2N+1, \end{cases} \end{aligned} \tag{4.29}$$

which allows us to evaluate the final term in Eq. (4.28).

Equations (4.27) and (4.28) specify the required $2N + 1$ equations for the $2N + 1$ unknown U_j, $j = 1, 2, \ldots, 2N + 1$, that appear in the expression for $U(x)$ given by Eq. (4.25). These equations may be written as the linear system

$$A\mathbf{U} = \mathbf{b}, \tag{4.30}$$

where, for $j = 1, 2, \ldots, 2N + 1$,

$$A_{1,j} = \begin{cases} 1, & j = 1, \\ 0, & j \neq 1, \end{cases} \tag{4.31}$$

$$A_{i,j} = \int_a^b p \frac{\mathrm{d}\phi_i}{\mathrm{d}x}\frac{\mathrm{d}\phi_j}{\mathrm{d}x} + q\phi_i \frac{\mathrm{d}\phi_j}{\mathrm{d}x} + r\phi_i\phi_j \,\mathrm{d}x, \qquad i = 2, 3, \ldots, 2N + 1, \tag{4.32}$$

and

$$b_1 = u_a, \tag{4.33}$$

$$b_i = \int_a^b f\phi_i \,\mathrm{d}x, \qquad i = 2, 3, \ldots, 2N, \tag{4.34}$$

$$b_{2N+1} = \int_a^b f\phi_{2N+1} \,\mathrm{d}x + \eta_b. \tag{4.35}$$

4.2.5 Assembling the Algebraic Equations

We will assemble the system of algebraic equations given by Eq. (4.30) in the same spirit as in Sect. 3.4, when linear basis functions were used. We first calculate entries of the form given by Eqs. (4.32), (4.34) and (4.35) that require the evaluation of integrals of known functions over the whole domain $a < x < b$. These integrals are evaluated by decomposing them into the sum of contributions from individual elements. Remembering that element k occupies the region $x_{2k-1} \leq x \leq x_{2k+1}$, we may write Eq. (4.32) as, for $i = 2, 3, \ldots, 2N + 1$, $j = 1, 2, \ldots, 2N + 1$:

$$A_{i,j} = \sum_{k=1}^{N} \int_{x_{2k-1}}^{x_{2k+1}} p \frac{\mathrm{d}\phi_i}{\mathrm{d}x}\frac{\mathrm{d}\phi_j}{\mathrm{d}x} + q\phi_i \frac{\mathrm{d}\phi_j}{\mathrm{d}x} + r\phi_i\phi_j \,\mathrm{d}x, \tag{4.36}$$

and Eqs. (4.34) and (4.35) as

$$b_i = \sum_{k=1}^{N} \int_{x_{2k-1}}^{x_{2k+1}} f\phi_i \,\mathrm{d}x, \qquad i = 2, 3, \ldots, 2N, \tag{4.37}$$

$$b_{2N+1} = \sum_{k=1}^{N} \int_{x_{2k-1}}^{x_{2k+1}} f\phi_{2N+1} \,\mathrm{d}x + \eta_b. \tag{4.38}$$

Having calculated these entries, we complete the assembly of the linear system by setting entries defined by Eqs. (4.31) and (4.33) that are given by explicit expressions. We now describe the assembly of both categories of entry in more detail.

4.2.5.1 The Contributions from Integrals

We now explain how to evaluate the contributions to Eqs. (4.36)–(4.38) from individual elements. When evaluating local contributions from individual elements in Sect. 3.4.1, when linear basis functions were used, we noted that only two basis functions were nonzero on each element. Nonzero contributions from element k to the global matrix A and global vector $\mathbf{b}$ were then stored in a local matrix $A_{\text{local}}^{(k)}$ of size 2×2, and a local vector $\mathbf{b}_{\text{local}}^{(k)}$ with two entries. We take a similar approach here.

The contributions from element k to Eq. (4.36) are given by, for $i = 2, 3, \ldots, 2N+1$, $j = 1, 2, \ldots, 2N+1$:

$$\int_{x_{2k-1}}^{x_{2k+1}} p\frac{\mathrm{d}\phi_i}{\mathrm{d}x}\frac{\mathrm{d}\phi_j}{\mathrm{d}x} + q\phi_i\frac{\mathrm{d}\phi_j}{\mathrm{d}x} + r\phi_i\phi_j \, \mathrm{d}x.$$

The only basis functions that are nonzero on element k are the three basis functions ϕ_{2k-1}, ϕ_{2k} and ϕ_{2k+1}. The integral above is therefore zero, unless both i and j take one of the values $2k-1$, $2k$ or $2k+1$. We may therefore store the nonzero local contributions to Eq. (4.36) from element k in the 3×3 matrix $A_{\text{local}}^{(k)}$. The relationship between the indices of the entries in this local matrix, and the indices of the entries in the global matrix A that they contribute to, is given in Table 4.1.

Similarly, the contributions from element k to Eqs. (4.37) and (4.38) are given by, for $i = 2, 3, \ldots, 2N+1$:

$$\sum_{k=1}^{N} \int_{x_{2k-1}}^{x_{2k+1}} f\phi_i \, \mathrm{d}x.$$

This integral is zero unless i takes one of the values $2k-1$, $2k$ or $2k+1$. We may therefore store the nonzero local contributions to Eqs. (4.37) and (4.38) from element k in the vector $\mathbf{b}_{\text{local}}^{(k)}$ with three entries. The relationship between the indices of the

Table 4.1 The relationship between the location of the entries of $A_{\text{local}}^{(k)}$ and $\mathbf{b}_{\text{local}}^{(k)}$, and the location of the entries of A and $\mathbf{b}$ that they contribute to, when quadratic basis functions are used

Local index	Global index
1	$2k-1$
2	$2k$
3	$2k+1$

entries in this local vector, and the indices of the entries in the global vector $\mathbf{b}$ that they contribute to, is given in Table 4.1.

The entries of $A_{\text{local}}^{(k)}$ and $\mathbf{b}_{\text{local}}^{(k)}$ are given below.

$$
\begin{aligned}
A_{\text{local},\,1,1}^{(k)} &= \int_{x_{2k-1}}^{x_{2k+1}} p\frac{\mathrm{d}\phi_{2k-1}}{\mathrm{d}x}\frac{\mathrm{d}\phi_{2k-1}}{\mathrm{d}x} + q\phi_{2k-1}\frac{\mathrm{d}\phi_{2k-1}}{\mathrm{d}x} + r\phi_{2k-1}\phi_{2k-1}\,\mathrm{d}x,\\
A_{\text{local},\,1,2}^{(k)} &= \int_{x_{2k-1}}^{x_{2k+1}} p\frac{\mathrm{d}\phi_{2k-1}}{\mathrm{d}x}\frac{\mathrm{d}\phi_{2k}}{\mathrm{d}x} + q\phi_{2k-1}\frac{\mathrm{d}\phi_{2k}}{\mathrm{d}x} + r\phi_{2k-1}\phi_{2k}\,\mathrm{d}x,\\
A_{\text{local},\,1,3}^{(k)} &= \int_{x_{2k-1}}^{x_{2k+1}} p\frac{\mathrm{d}\phi_{2k-1}}{\mathrm{d}x}\frac{\mathrm{d}\phi_{2k+1}}{\mathrm{d}x} + q\phi_{2k-1}\frac{\mathrm{d}\phi_{2k+1}}{\mathrm{d}x} + r\phi_{2k-1}\phi_{2k+1}\,\mathrm{d}x,\\
A_{\text{local},\,2,1}^{(k)} &= \int_{x_{2k-1}}^{x_{2k+1}} p\frac{\mathrm{d}\phi_{2k}}{\mathrm{d}x}\frac{\mathrm{d}\phi_{2k-1}}{\mathrm{d}x} + q\phi_{2k}\frac{\mathrm{d}\phi_{2k-1}}{\mathrm{d}x} + r\phi_{2k}\phi_{2k-1}\,\mathrm{d}x,\\
A_{\text{local},\,2,2}^{(k)} &= \int_{x_{2k-1}}^{x_{2k+1}} p\frac{\mathrm{d}\phi_{2k}}{\mathrm{d}x}\frac{\mathrm{d}\phi_{2k}}{\mathrm{d}x} + q\phi_{2k}\frac{\mathrm{d}\phi_{2k}}{\mathrm{d}x} + r\phi_{2k}\phi_{2k}\,\mathrm{d}x,\\
A_{\text{local},\,2,3}^{(k)} &= \int_{x_{2k-1}}^{x_{2k+1}} p\frac{\mathrm{d}\phi_{2k}}{\mathrm{d}x}\frac{\mathrm{d}\phi_{2k+1}}{\mathrm{d}x} + q\phi_{2k}\frac{\mathrm{d}\phi_{2k+1}}{\mathrm{d}x} + r\phi_{2k}\phi_{2k+1}\,\mathrm{d}x,\\
A_{\text{local},\,3,1}^{(k)} &= \int_{x_{2k-1}}^{x_{2k+1}} p\frac{\mathrm{d}\phi_{2k+1}}{\mathrm{d}x}\frac{\mathrm{d}\phi_{2k-1}}{\mathrm{d}x} + q\phi_{2k+1}\frac{\mathrm{d}\phi_{2k-1}}{\mathrm{d}x} + r\phi_{2k+1}\phi_{2k-1}\,\mathrm{d}x,\\
A_{\text{local},\,3,2}^{(k)} &= \int_{x_{2k-1}}^{x_{2k+1}} p\frac{\mathrm{d}\phi_{2k+1}}{\mathrm{d}x}\frac{\mathrm{d}\phi_{2k}}{\mathrm{d}x} + q\phi_{2k+1}\frac{\mathrm{d}\phi_{2k}}{\mathrm{d}x} + r\phi_{2k+1}\phi_{2k}\,\mathrm{d}x,\\
A_{\text{local},\,3,3}^{(k)} &= \int_{x_{2k-1}}^{x_{2k+1}} p\frac{\mathrm{d}\phi_{2k+1}}{\mathrm{d}x}\frac{\mathrm{d}\phi_{2k+1}}{\mathrm{d}x} + q\phi_{2k+1}\frac{\mathrm{d}\phi_{2k+1}}{\mathrm{d}x} + r\phi_{2k+1}\phi_{2k+1}\,\mathrm{d}x,\\
b_{\text{local},\,1}^{(k)} &= \int_{x_{2k-1}}^{x_{2k+1}} f\phi_{2k-1}\,\mathrm{d}x,\\
b_{\text{local},\,2}^{(k)} &= \int_{x_{2k-1}}^{x_{2k+1}} f\phi_{2k}\,\mathrm{d}x,\\
b_{\text{local},\,3}^{(k)} &= \int_{x_{2k-1}}^{x_{2k+1}} f\phi_{2k+1}\,\mathrm{d}x.
\end{aligned}
$$

As with the integrals over each element that arose in Chap. 3 when linear basis functions were used, there is no guarantee that the integrals above may be evaluated analytically. Even if analytic evaluation is possible, it may require much tedious, error-prone algebra. As such, it is often necessary, or desirable, to approximate these integrals using numerical quadrature, as described in Sect. 3.5. When approximating these integrals over element k using quadrature, we will use the change of variable given by Eq. (4.6), which we now write as

$$x = x_{2k-1} + hX, \qquad 0 \leq X \leq 1, \tag{4.39}$$

where $h = x_{2k+1} - x_{2k-1}$. Using the change of variable given by Eq. (4.39), and the definition of global basis functions given by Eqs. (4.6)–(4.13), the entries of $A^{(k)}_{\text{local}}$ and $\mathbf{b}^{(k)}_{\text{local}}$ may be written as

$$A^{(k)}_{\text{local},\,i,j} = h\int_0^1 \frac{1}{h^2}p(x_{2k-1}+hX)\frac{\mathrm{d}\phi_{\text{local},\,i}}{\mathrm{d}X}\frac{\mathrm{d}\phi_{\text{local},\,j}}{\mathrm{d}X} + \frac{1}{h}q(x_{2k-1}+hX)\phi_{\text{local},\,i}(X)\frac{\mathrm{d}\phi_{\text{local},\,j}}{\mathrm{d}X} + r(x_{2k-1}+hX)\phi_{\text{local},\,i}(X)\phi_{\text{local},\,j}(X)\,\mathrm{d}X, \qquad i,j=1,2,3, \tag{4.40}$$

$$b^{(k)}_{\text{local},\,i} = h\int_0^1 f(x_{2k-1}+hX)\phi_{\text{local},\,i}(X)\,\mathrm{d}X, \qquad i=1,2,3. \tag{4.41}$$

The integration required is now over the interval $0 \le X \le 1$, allowing us to approximate these integrals using the quadrature techniques described in Sect. 3.5. Note the similarity of Eqs. (4.40) and (4.41) with Eqs. (3.59) and (3.60). The only differences are the range of the indices i and j, and the definition of local basis functions.

The entries of A and $\mathbf{b}$ that are given by Eqs. (4.32), (4.34) and (4.35) may now be calculated by looping over all elements, calculating the nonzero local contributions from each element given by $A^{(k)}_{\text{local}}$ and $\mathbf{b}^{(\mathbf{k})}_{\text{local}}$, and then adding these contributions to the correct entries of A and $\mathbf{b}$ as given in Table 4.1. Finally, we complete the assembly of b_{2N+1} by adding the final term on the right hand side of Eq. (4.35) to this entry of $\mathbf{b}$.

4.2.5.2 Completing the Assembly of the Linear System

The final task when assembling the system of algebraic equations given by Eq. (4.30) is to set the entries that are given by Eqs. (4.31) and (4.33). We explained in Sect. 2.8.2 that, when computing the entries of the linear system that are given by equations such as Eqs. (4.32), (4.34) and (4.35) using the method described in Sect. 4.2.5.1, incorrect entries would be entered into the rows of the linear system where equations such as Eqs. (4.31) and (4.33) apply. We therefore have to ensure that these entries are overwritten by the correct entries.

4.2.6 Computational Implementation

We now provide an exemplar MATLAB implementation of the material presented in this chapter, using the same boundary value problem that was used in Chap. 3 which we re-state here. We calculate a finite element solution, using a quadratic approximation on each element, of the differential equation

$$-\frac{\mathrm{d}}{\mathrm{d}x}\left(-x\frac{\mathrm{d}u}{\mathrm{d}x}\right)-(2x+3)\frac{\mathrm{d}u}{\mathrm{d}x}+(x+2)u=x^3\mathrm{e}^{2x}, \qquad 0.5<x<1, \tag{4.42}$$

together with a Dirichlet boundary condition at $x = 0.5$:

$$u=\mathrm{e}, \qquad x=0.5, \tag{4.43}$$

and a natural Neumann boundary condition at $x = 1$:

$$-x\frac{\mathrm{d}u}{\mathrm{d}x}=-5\mathrm{e}^2, \qquad x=1. \tag{4.44}$$

This boundary value problem has solution

$$u(x)=\mathrm{e}^x\left(K_1+K_2x^3\right)+\mathrm{e}^{2x}\left(x^2-2x+2\right),$$

where

$$K_1 = 3\mathrm{e}-\frac{32}{31}\left(3\mathrm{e}+\frac{\sqrt{\mathrm{e}}}{4}\right),$$
$$K_2 = \frac{8}{31}\left(3\mathrm{e}+\frac{\sqrt{\mathrm{e}}}{4}\right).$$

This boundary value problem may be related to the model problem defined by Eqs. (4.1)–(4.3) by writing

$$\begin{aligned}
p(x) &= -x,\\
q(x) &= -(2x+3),\\
r(x) &= x+2,\\
f(x) &= x^3\mathrm{e}^{2x},\\
u_a &= \mathrm{e},\\
\eta_b &= -5\mathrm{e}^2.
\end{aligned}$$

A MATLAB implementation is given in Listing 4.1. This listing uses the same pattern as in earlier chapters, where variable names follow the notation used in this chapter.

The MATLAB function requires an array `x_ele_bound` as input. This array should contain only the nodes on the boundary of elements, and the entries should be listed in ascending order. Thus, if N elements are used, the array should be of length $N + 1$. The first entry should take the value 0.5, i.e. the value of x at the left hand boundary, and the last entry should take the value 1, i.e. the value of x at the right hand boundary. The function then returns a vector $\mathbf{x}$ that contains all nodes in the mesh, and a vector $\mathbf{U}$ that contains the finite element solution at each node. The function may be called by typing, for example,

```
x_ele_bound = linspace(0.5,1,101);
[x, U] = Chap4_CalculateQuadraticFemForBvp(x_ele_bound);
```

into a MATLAB session. This will calculate the finite element solution on a mesh with 100 equally sized elements.

The functions $p(x), q(x), r(x), f(x)$ and the values u_a, η_b are defined in lines 5–10 of the listing. Noting that the vector `x_ele_bound` contains $N + 1$ nodes, where N is the number of elements, we set the number of elements in line 13. There will be $2N + 1$ nodes in the mesh, and so we allocate the appropriate memory for the vector of nodes $\mathbf{x}$ in line 16. The nodes x_{2k-1}, $k = 1, 2, \ldots, N + 1$, are the entries of the vector `x_ele_bound`—these are set in line 19. The nodes x_{2k}, $k = 1, 2, \ldots, N$, are the midpoints of the elements—these are set in line 22. We then initialise the entries of the global matrix A and vector $\mathbf{b}$ to zero in lines 25 and 26, before looping over all elements in lines 30–37 to first calculate the local contributions to A and $\mathbf{b}$, and then increment A and $\mathbf{b}$ appropriately.

The local contributions to both A and $\mathbf{b}$ from each element are calculated using the function `CalculateContributionOnElement` given by lines 58–102 of the listing. This function first calculates the length of the interval in line 62 before setting the number of Gaussian quadrature points, the quadrature points and the quadrature weights in lines 65–67. The entries of the local contributions to A and $\mathbf{b}$ are initialised to zero in lines 70–71. We then loop over all Gaussian quadrature points in lines 74–102. At each Gauss point, we calculate the value of all functions appearing in Eqs. (4.40) and (4.41) in lines 78–85. We then loop over all entries of the nonzero local contributions to the linear system, incrementing each entry with the appropriate weighting from the Gaussian quadrature scheme.

Returning to the main function, we handle the Neumann boundary conditions in line 41, and the Dirichlet boundary conditions using lines 44–46. Finally, we solve the linear system in line 49, before plotting the solution using lines 52–54. Note that the solution plotted is not exactly the finite element solution—line 52 of the code simply joins successive points in the mesh using a straight line, rather than plotting the quadratic approximation to the solution on each element. We leave plotting the finite element solution as a quadratic approximation on each element as an exercise for the reader.

Listing 4.1 `Chap4_CalculateQuadraticFemForBvp.m`, a MATLAB function for calculating the finite element solution of a boundary value problem of the form given in Sect. 4.1.

```
1  function [x, U] = ...
2      Chap4_CalculateQuadraticFemForBvp(x_ele_bound)
3
4  % define p(x), q(x), r(x), f(x), u_a, eta_b
5  p = @(x) -x;
6  q = @(x) -(2*x+3);
7  r = @(x) x+2;
8  f = @(x) x^3*exp(2*x);
9  u_a = exp(1);
10 eta_b = -5*exp(1)*exp(1);
11
12 % Deduce number of elements
```

```
N = length(x_ele_bound)-1;

% Allocate memory for nodes
x = zeros(2*N+1, 1);

% Define nodes on edge of elements
x(1:2:2*N+1) = x_ele_bound;

% Define nodes at centre of elements
x(2:2:2*N) = 0.5*(x(1:2:2*N-1) + x(3:2:2*N+1));

% Initialise A and b to zero
A = zeros(2*N+1, 2*N+1);
b = zeros(2*N+1, 1);

% Loop over elements calculating local contributions and
% inserting into linear system
for k = 1:N
    [A_local, b_local] = ...
        CalculateContributionOnElement(x, k, p, q, r, f);

    A(2*k-1:2*k+1, 2*k-1:2*k+1) =  ...
        A(2*k-1:2*k+1, 2*k-1:2*k+1) + A_local;
    b(2*k-1:2*k+1) =  b(2*k-1:2*k+1) + b_local;
end

% Add contributions from Neumann boundary conditions into
% linear system
b(2*N+1) = b(2*N+1) + eta_b;

% Set Dirichlet boundary conditions in row 1
A(1,:) = 0;
A(1,1) = 1;
b(1) = u_a;

% Solve linear system
U = A\b;

% Plot finite element solution
plot(x,U,'-o')
xlabel('x')
ylabel('U')



function [A_local, b_local] = ...
    CalculateContributionOnElement(x, k, p, q, r, f)

% Calculate element length
h = x(2*k+1) - x(2*k-1);

% Define quadrature rule
M = 3;
QuadWeights = [5/18, 4/9, 5/18];
QuadPoints = [0.5*(1-sqrt(0.6)), 0.5, 0.5*(1+sqrt(0.6))];

% Initialise local contributions to A and b to zero
A_local = zeros(3,3);
b_local = zeros(3,1);

% Loop over quadrature points
for m = 1:M
    X = QuadPoints(m);
```

```
    % Evaluate local basis functions and their derivatives
    phi_local = [2*(1-X)*(0.5-X), 4*X*(1-X), 2*X*(X-0.5)];
    phi_prime_local = [4*X-3, 4-8*X, 4*X-1];

    %Evaluate p, q, r, f at quadrature points
    p_val = p(x(2*k-1)+h*X);
    q_val = q(x(2*k-1)+h*X);
    r_val = r(x(2*k-1)+h*X);
    f_val = f(x(2*k-1)+h*X);

    % Increment entries of A_local and b_local
    % with suitably weighted function evaluations
    for i=1:3
        for j=1:3
            A_local(i,j) = A_local(i,j) + ...
                h*QuadWeights(m)*( ...
                p_val/h/h*phi_prime_local(i)* ...
                phi_prime_local(j) + ...
                q_val/h*phi_local(i)* ...
                phi_prime_local(j) + ...
                r_val*phi_local(i)*phi_local(j));
        end
        b_local(i) = b_local(i) + h*QuadWeights(m)* ...
            f_val*phi_local(i);
    end
end
```

4.2.7 A Note on Using Quadrature for Higher Order Basis Functions

We have seen that the assembly of the system of algebraic equations that determines the finite element solution requires the evaluation of many integrals over the whole computational domain. Throughout this book, we evaluate the integrals by first calculating the nonzero contributions from each element, before adding these local contributions into the global matrix A and global vector $\mathbf{b}$. These integrals may not, in general, be evaluated analytically and are usually computed using Gaussian quadrature. It should be noted that increasing the order of the approximating polynomial on each element increases the overall degree of the function that is being integrated. As a consequence, more quadrature points may be needed to evaluate these integrals sufficiently accurately.

4.3 Cubic Basis Functions

In Chap. 3, we explained how to calculate the finite element solution of a differential equation, using a linear approximation to the solution on each element. In Sect. 4.2, we developed the finite element method further, making a quadratic approximation to the solution on each element. In this section, we explain, albeit in less detail, how to calculate a finite element solution that is a cubic approximation to the solution on each element. The key ingredients required that we do not already have are a suitable computational mesh, and cubic basis functions defined on this mesh. We again use the boundary value problem given by Eqs. (4.1)–(4.3) when explaining how cubic basis functions may be used.

4.3.1 The Mesh

We again partition the computational domain $a < x < b$ into N elements. A cubic approximation to the solution on each element requires that every element contains four nodes. Two of these nodes will be the end-points of these elements. The other two nodes—the *interior nodes*—will be placed inside the element so that the four nodes are uniformly spaced across the element. There will be a total of $N + 1$ nodes at the end-points of the N elements, and a further $2N$ nodes in the interior of the elements, giving a total of $3N + 1$ nodes.

The nodes are numbered as follows. We demand that: (i) $x_1 = a$; (ii) $x_{3N+1} = b$; (iii) the nodes are listed in ascending order, so that $x_{j+1} > x_j$ for $j = 1, 2, \dots, 3N$; and (iv) the nodes contained in each element are uniformly spaced across the element. Using this convention, element k, where $k = 1, 2, \dots, N$, occupies the region $x_{3k-2} \le x \le x_{3k+1}$, and the elements span the whole of the computational domain $a < x < b$. Insisting that the interior nodes of element k, where $k = 1, 2, \dots, N$, are uniformly spaced across the element implies that:

$$x_{3k-1} = \frac{1}{3}\left(2x_{3k-2} + x_{3k+1}\right), \tag{4.45}$$

$$x_{3k} = \frac{1}{3}\left(x_{3k-2} + 2x_{3k+1}\right). \tag{4.46}$$

An example mesh, that may be used to calculate a finite element solution that is a cubic approximation to the solution on each element, is illustrated in Fig. 4.4.

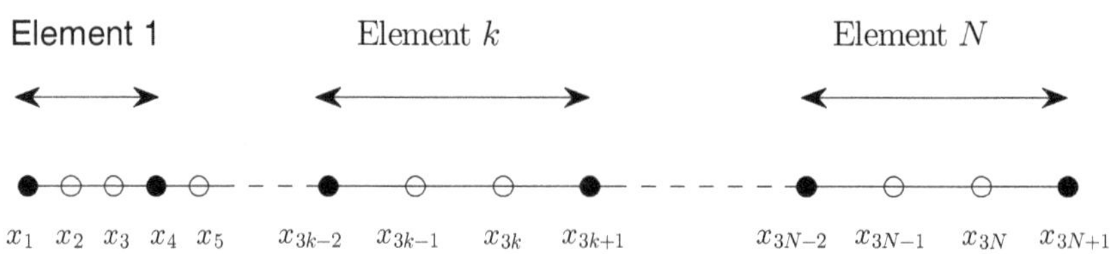

Fig. 4.4 The nodes and elements of a finite element mesh that may be used to calculate a finite element solution that is a cubic approximation on each element. Nodes on the boundary of an element are denoted by a *filled circle*, and nodes in the interior of an element are denoted by an *open circle*

4.3.2 *The Definition of Cubic Basis Functions*

The quadratic basis functions used in Sect. 4.2 were defined in terms of local basis functions defined on the canonical element $0 \leq X \leq 1$. This approach may be adapted to allow cubic basis functions to be defined. We first need a mapping between element k of the mesh and the canonical element $0 \leq X \leq 1$. As element k, where $k = 1, 2, \ldots, N$, occupies the region $x_{3k-2} \leq x \leq x_{3k+1}$, we may map between this element and the canonical element using the change of variable

$$x = x_{3k-2} + (x_{3k+1} - x_{3k-2})X, \qquad 0 \leq X \leq 1. \tag{4.47}$$

Local basis functions are then given by

$$\phi_{\text{local},1}(X) = \frac{9}{2}\left(\frac{1}{3} - X\right)\left(\frac{2}{3} - X\right)(1 - X), \tag{4.48}$$

$$\phi_{\text{local},2}(X) = \frac{27}{2}X\left(\frac{2}{3} - X\right)(1 - X), \tag{4.49}$$

$$\phi_{\text{local},3}(X) = \frac{27}{2}X\left(X - \frac{1}{3}\right)(1 - X), \tag{4.50}$$

$$\phi_{\text{local},4}(X) = \frac{9}{2}X\left(X - \frac{1}{3}\right)\left(X - \frac{2}{3}\right). \tag{4.51}$$

These local basis functions are illustrated in Fig. 4.5. Global basis functions $\phi_j(x)$, $j = 1, 2, \ldots, 3N + 1$, may now be defined using these local basis functions. On element k, we define the global basis functions in terms of the local basis functions by:

$$\phi_{3k-2}(X) = \phi_{\text{local},1}(X), \qquad 0 \leq X \leq 1, \tag{4.52}$$

$$\phi_{3k-1}(X) = \phi_{\text{local},2}(X), \qquad 0 \leq X \leq 1, \tag{4.53}$$

$$\phi_{3k}(X) = \phi_{\text{local},3}(X), \qquad 0 \leq X \leq 1, \tag{4.54}$$

$$\phi_{3k+1}(X) = \phi_{\text{local},4}(X), \qquad 0 \leq X \leq 1, \tag{4.55}$$

$$\phi_j(X) = 0, \qquad 0 \leq X \leq 1, \qquad j \neq 3k - 2, 3k - 1, 3k, 3k + 1. \tag{4.56}$$

The basis functions $\phi_{3k-2}(x)$, $\phi_{3k-1}(x)$, $\phi_{3k}(x)$, $\phi_{3k+1}(x)$ are shown in Fig. 4.6.

Using the global basis functions given by Eqs. (4.47)–(4.56), and remembering the location of interior nodes given by Eqs. (4.45) and (4.46), we may deduce the following properties of these global basis functions on element k:

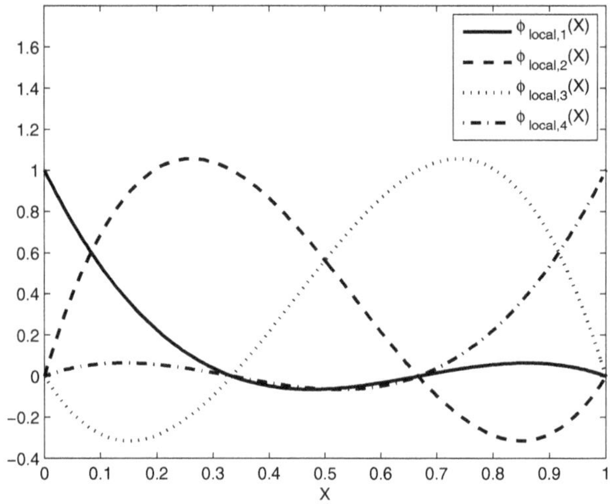

Fig. 4.5 The local basis functions $\phi_{\text{local},1}(X)$, $\phi_{\text{local},2}(X)$, $\phi_{\text{local},3}(X)$ and $\phi_{\text{local},4}(X)$ defined by Eqs. (4.48)–(4.51)

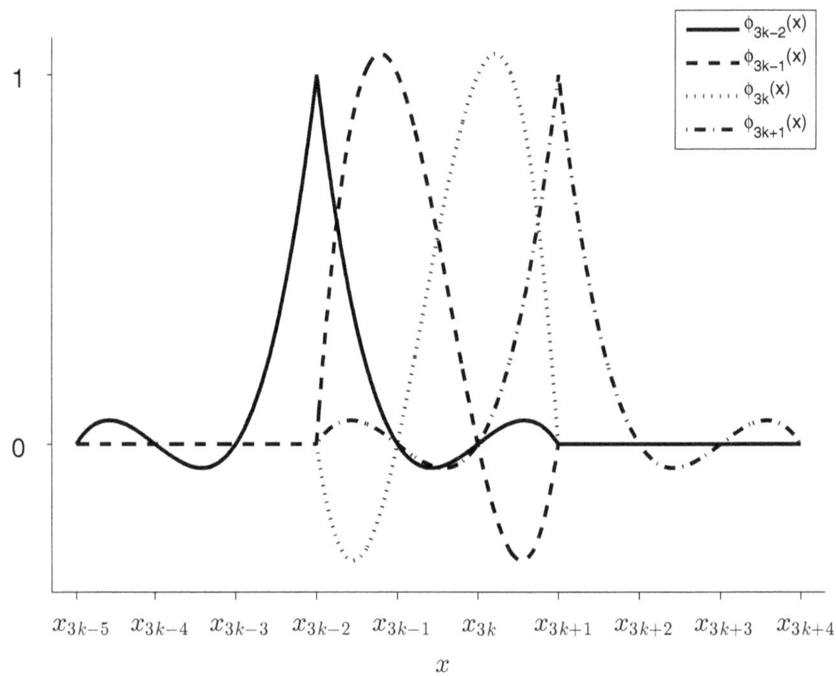

Fig. 4.6 The basis functions $\phi_{3k-2}(x)$, $\phi_{3k-1}(x)$, $\phi_{3k}(x)$ and $\phi_{3k+1}(x)$ defined by Eqs. (4.47)–(4.56)

$$\begin{array}{llll} \phi_{3k-2}(x_{3k-2}) = 1, & \phi_{3k-2}(x_{3k-1}) = 0, & \phi_{3k-2}(x_{3k}) = 0, & \phi_{3k-2}(x_{3k+1}) = 0, \\ \phi_{3k-1}(x_{3k-2}) = 0, & \phi_{3k-1}(x_{3k-1}) = 1, & \phi_{3k-1}(x_{3k}) = 0, & \phi_{3k-1}(x_{3k+1}) = 0, \\ \phi_{3k}(x_{3k-2}) = 0, & \phi_{3k}(x_{3k-1}) = 0, & \phi_{3k}(x_{3k}) = 1, & \phi_{3k}(x_{3k+1}) = 0, \\ \phi_{3k+1}(x_{3k-2}) = 0, & \phi_{3k+1}(x_{3k-1}) = 0, & \phi_{3k+1}(x_{3k}) = 0, & \phi_{3k+1}(x_{3k+1}) = 1, \end{array}$$

and, for $j \neq 3k-2, 3k-1, 3k, 3k+1$,

$$\phi_j(x_{3k-2}) = 0, \qquad \phi_j(x_{3k-1}) = 0, \qquad \phi_j(x_{3k}) = 0, \qquad \phi_j(x_{3k+1}) = 0.$$

These properties of basis functions hold on all elements. Hence, these basis functions satisfy the usual property that, for $j = 1, 2, \ldots, 3N+1$,

$$\phi_j(x_i) = \begin{cases} 1, & i = j, \\ 0, & i \neq j. \end{cases} \tag{4.57}$$

4.3.3 The Finite Element Solution

We now explain how the cubic basis functions, defined in Sect. 4.3.2, may be used to calculate the finite element solution of the boundary value problem given by Eqs. (4.1)–(4.3). The set S^h is now defined to be those functions $V(x)$ that can be written as

$$V(x) = \sum_{j=1}^{3N+1} V_j \phi_j(x),$$

where $\phi_j(x)$, $j = 1, 2, \ldots, 3N+1$, are defined by Eqs. (4.47)–(4.56), with the subsets S_E^h and S_0^h of the set S^h defined as before. The finite element solution $U(x)$ is defined, as in earlier examples, by

find $U(x) \in S_E^h$ such that

$$\int_a^b p \frac{\mathrm{d}U}{\mathrm{d}x} \frac{\mathrm{d}\phi_i}{\mathrm{d}x} + q \frac{\mathrm{d}U}{\mathrm{d}x} \phi_i + rU\phi_i \, \mathrm{d}x = \int_a^b f\phi_i \, \mathrm{d}x + \eta_b \phi_i(b), \tag{4.58}$$

for all $\phi_i(x) \in S_0^h$.

Writing the finite element solution as the member of S^h given by

$$U(x) = \sum_{j=1}^{3N+1} U_j \phi_j(x), \tag{4.59}$$

the specification of the finite element solution by Eq. (4.58) allows us to derive a system of algebraic equations that determines the unknown values $U_1, U_2, \ldots, U_{3N+1}$. As usual, these equations are drawn from two sources: (i) equations that ensure that the finite element solution is in the set S_E^h, i.e. we insist that $U(x)$ satisfies the Dirichlet boundary conditions and (ii) the substitution of suitable test functions $\phi_i(x) \in S_0^h$ into Eq. (4.58).

There is one Dirichlet boundary condition, given by Eq. (4.2), and it is applied at the node where $x = x_1$. Using Eqs. (4.57) and (4.59), we see that this equation is satisfied if

$$U_1 = u_a. \tag{4.60}$$

Noting that $\phi_i(x) \in S_0^h$, for $i = 2, 3, \ldots, 3N + 1$, we obtain a further $3N$ algebraic equations by substituting these test functions into Eq. (4.58). Combined with Eq. (4.60), we now have $3N + 1$ equations, as would be expected to determine $3N + 1$ unknown values $U_1, U_2, \ldots, U_{3N+1}$. Using similar manipulations to that used for quadratic basis functions in Sect. 4.2.4, we may write these equations as

$$A\mathbf{U} = \mathbf{b}, \tag{4.61}$$

where, for $j = 1, 2, \ldots, 3N + 1$,

$$A_{1,j} = \begin{cases} 1, & j = 1, \\ 0, & j \neq 1, \end{cases} \tag{4.62}$$

$$A_{i,j} = \int_a^b p\frac{\mathrm{d}\phi_i}{\mathrm{d}x}\frac{\mathrm{d}\phi_j}{\mathrm{d}x} + q\phi_i\frac{\mathrm{d}\phi_j}{\mathrm{d}x} + r\phi_i\phi_j \,\mathrm{d}x, \qquad i = 2, 3, \ldots, 3N + 1, \tag{4.63}$$

and

$$b_1 = u_a, \tag{4.64}$$

$$b_i = \int_a^b f\phi_i \,\mathrm{d}x, \qquad i = 2, 3, \ldots, 3N, \tag{4.65}$$

$$b_{3N+1} = \int_a^b f\phi_{2N+1} \,\mathrm{d}x + \eta_b. \tag{4.66}$$

4.3.4 Assembling the Algebraic Equations

We now give a very brief account of how the linear system given by Eq. (4.61) may be assembled. We use the same strategy as in previous examples, and first compute the entries, given by Eqs. (4.63), (4.65) and (4.66), that are defined by integrals over the whole domain. Having computed these entries, we then complete the assembly of A and $\mathbf{b}$ by setting the entries given by Eqs. (4.62) and (4.64).

Table 4.2 The relation between the location of the entries in $A_{\text{local}}^{(k)}$ and $\mathbf{b}_{\text{local}}^{(k)}$, and the location of the entries of A and $\mathbf{b}$ that they contribute to, when cubic basis functions are used

Local index	Global index
1	$3k-2$
2	$3k-1$
3	$3k$
4	$3k+1$

As in earlier examples, the entries of the linear system given by Eqs. (4.63), (4.65) and (4.66) may be calculated efficiently by decomposing them into the sum of the nonzero contributions from integrals over individual elements. We note that, on element k, there are only four nonzero basis functions: $\phi_{3k-2}(x)$, $\phi_{3k-1}(x)$, $\phi_{3k}(x)$, $\phi_{3k+1}(x)$. Using the relation between local and global indices given in Table 4.2, we may store the nonzero contributions from element k to Eqs. (4.63), (4.65) and (4.66) in a 4×4 local matrix $A_{\text{local}}^{(k)}$, and a local vector $\mathbf{b}_{\text{local}}^{(k)}$ with four entries. These local contributions are given in terms of integrals over the canonical element by

$$
\begin{aligned}
A_{\text{local},\, i,j}^{(k)} &= h \int_0^1 \frac{1}{h^2} p(x_{3k-2}+hX) \frac{\mathrm{d}\phi_{\text{local},\, i}}{\mathrm{d}X} \frac{\mathrm{d}\phi_{\text{local},\, j}}{\mathrm{d}X} + \\
&\quad \frac{1}{h} q((x_{3k-2}+hX)\phi_{\text{local},\, i}(x) \frac{\mathrm{d}\phi_{\text{local},\, j}}{\mathrm{d}X} + \\
&\quad r((x_{3k-2}+hX)\phi_{\text{local},\, i}(X)\phi_{\text{local},\, j}(X)\, \mathrm{d}X, \quad i, j = 1, 2, 3, 4, \qquad (4.67) \\
b_{\text{local},\, i}^{(k)} &= h \int_0^1 f(x_{3k-2}+hX)\phi_{\text{local},\, i}(X)\, \mathrm{d}X, \qquad i = 1, 2, 3, 4, \qquad (4.68)
\end{aligned}
$$

where $h = x_{3k+1} - x_{3k-2}$, and the local basis functions are defined by Eqs. (4.47)–(4.56).

Entries of the linear system given by Eqs. (4.63), (4.65) and (4.66) may therefore be calculated by looping over all elements, calculating the nonzero local contributions from each element given by Eqs. (4.67) and (4.68), and adding these local contributions into the global matrix A and global vector $\mathbf{b}$. We then add the final term required to assemble Eq. (4.66), before setting the entries given by Eqs. (4.62) and (4.64).

4.4 General Basis Functions

The material presented in Sect. 4.2 on quadratic basis functions, and in Sect. 4.3 on cubic basis functions, may easily be generalised to a polynomial approximation of an arbitrary degree. We now give a very brief overview of how this may be done.

4.4.1 The Mesh

Suppose, we use a polynomial approximation to the solution on each element that is of degree α, where $\alpha \geq 1$, and partition the computational domain $a < x < b$ into N elements. A total of $\alpha + 1$ nodes will be required on each element to provide a polynomial approximation of degree α. On each element, there will be one node at either end of the element. We therefore need an extra $\alpha - 1$ interior nodes on each element. This gives a total of $\alpha N + 1$ nodes in the mesh.

We label the nodes so that they are ordered in ascending order and cover the whole domain. There are several consequences to these conditions. First, $x_1 = a$ and $x_{\alpha N+1} = b$. Secondly, as the nodes are listed in ascending order we have $x_{j+1} > x_j$ for $j = 1, 2, \ldots, \alpha N$. Finally, element k, where $k = 1, 2, \ldots, N$, occupies the region $x_{\alpha(k-1)+1} \leq x \leq x_{\alpha k+1}$.

We now define the $\alpha - 1$ interior nodes of element k, where $k = 1, 2, \ldots, N$, to be uniformly spaced across the element. The location of these nodes is then given by:

$$x_{\alpha(k-1)+j+1} = \frac{1}{\alpha}\left((\alpha - j)x_{\alpha(k-1)+1} + jx_{\alpha k+1}\right), \qquad j = 1, 2, \ldots, \alpha - 1.$$

4.4.2 Basis Functions

Global basis functions are again defined by: (i) defining a transformation between element k and the canonical element $0 \leq X \leq 1$; (ii) defining local basis functions on the canonical element; and (iii) defining the global basis functions in terms of the transformation and local basis functions.

Element k, where $k = 1, 2, \ldots, N$, occupies the region $x_{\alpha(k-1)+1} \leq x \leq x_{\alpha k+1}$. A transformation between element k and the canonical element $0 \leq X \leq 1$ is given by

$$x = x_{\alpha(k-1)+1} + \left(x_{\alpha k+1} - x_{\alpha(k-1)+1}\right) X, \qquad 0 \leq X \leq 1. \tag{4.69}$$

Local basis functions on the canonical element are given by

$$\phi_{\text{local},j}(X) = \frac{\Pi_{i \neq j}\,(X - X_i)}{\Pi_{i \neq j}\,\left(X_j - X_i\right)}, \qquad j = 1, 2, \ldots, \alpha + 1, \tag{4.70}$$

where

$$X_i = \frac{i-1}{\alpha}, \qquad i = 1, 2, \ldots, \alpha + 1,$$

and the expression $\Pi_{i \neq j}\,(X - X_i)$ is defined by

$$\Pi_{i\neq j}\left(X-X_i\right)=\left(X-X_1\right)\left(X-X_2\right)\dots\left(X-X_{j-1}\right)\left(X-X_{j+1}\right)\dots\left(X-X_{\alpha+1}\right).$$

It then follows that, for $j = 1, 2, \dots, \alpha + 1$,

$$\phi_{\text{local},j}(X_i)=\begin{cases}1, & i=j,\\ 0, & i\neq j.\end{cases} \tag{4.71}$$

Global basis functions on element k are then defined by

$$\phi_{\alpha(k-1)+j}(X)=\phi_{\text{local},j}(X), \qquad j=1,2,\dots,\alpha+1, \tag{4.72}$$
$$\phi_j(X)=0, \qquad \text{otherwise}. \tag{4.73}$$

Using this definition of global basis functions, we may use Eq. (4.71) to deduce that, for $j = 1, 2, \dots, \alpha N + 1$:

$$\phi_j(x_i)=\begin{cases}1, & i=j,\\ 0, & i\neq j.\end{cases}$$

4.4.3 The Finite Element Solution

The set S^h is now defined to be those functions $V(x)$ that can be written as

$$V(x)=\sum_{j=1}^{\alpha N+1} V_j\phi_j(x),$$

where $\phi_j(x)$, $j = 1, 2, \dots, \alpha N + 1$ are defined by Eqs. (4.69)–(4.73), with the subsets S_E^h and S_0^h of the set S^h defined as before. The finite element solution is then defined by

find $U(x) \in S_E^h$ such that

$$\int_a^b p\frac{\mathrm{d}U}{\mathrm{d}x}\frac{\mathrm{d}\phi_i}{\mathrm{d}x}+q\frac{\mathrm{d}U}{\mathrm{d}x}\phi_i+rU\phi_i\,\mathrm{d}x=\int_a^b f\phi_i\,\mathrm{d}x+\eta_b\phi_i(b), \tag{4.74}$$

for all $\phi_i(x) \in S_0^h$.

Writing the finite element solution as the member of the set S^h given by

$$U(x)=\sum_{j=1}^{\alpha N+1} U_j\phi_j(x),$$

a system of algebraic equations may be derived, in a similar manner to earlier examples, for the unknown values $U_1, U_2, \ldots, U_{\alpha N+1}$. These equations will again arise from two sources: equations that ensure that the finite element solution satisfies the Dirichlet boundary conditions, and equations that arise from substituting a suitable test function $\phi_i(x)$ into Eq. (4.74). These algebraic equations may be efficiently assembled by first evaluating the entries that are given by integrals over the whole domain; these entries are conveniently calculated by summing the contributions from individual elements. Having done this, the contributions from both Neumann and Dirichlet boundary conditions must be handled appropriately to complete the assembly of the linear system.

4.5 Convergence of the Finite Element Solution

We conclude this chapter by presenting a bound on the L^2 norm of the error of the finite element solution when a general polynomial approximation is made to the solution on each element.

Theorem 4.1 *Let $p(x), q(x), r(x), f(x)$ be functions such that $p(x), q(x), r(x)$, $f(x)$ and $\frac{\mathrm{d}p}{\mathrm{d}x}$ are continuous functions on the interval $a < x < b$. Suppose the second order differential equation, defined on $a < x < b$, and given by*

$$-\frac{\mathrm{d}}{\mathrm{d}x}\left(p(x)\frac{\mathrm{d}u}{\mathrm{d}x}\right) + q(x)\frac{\mathrm{d}u}{\mathrm{d}x} + r(x)u = f(x),$$

has a unique solution, subject to suitable given Dirichlet, Neumann or Robin boundary conditions. Let $U(x)$ be the finite element solution, calculated on a uniform mesh of elements of length h, using an approximating polynomial of degree α to the solution on each element. Define the error, $E(x)$, by

$$E(x) = U(x) - u(x),$$

and the L^2 norm of the error by

$$\|E\|_{L^2} = \sqrt{\int_a^b (E(x))^2 \,\mathrm{d}x}.$$

There then exists a constant $C > 0$, independent of the mesh, such that, as the element size $h \to 0$, the error satisfies

$$\|E\|_{L^2} = Ch^{\alpha+1}.$$

We see, from the theorem above, that if we increase the degree of the approximating polynomial by one, then the exponent of h in the error bound is increased by one. We also note that Theorem 3.1 is a special case of the theorem above, using the value $\alpha = 1$.

4.6 Exercises

The exercises below are all based on the same boundary value problem as Exercise 3.1 from Chap. 3, with parameters $\alpha = 5$ and $\beta = 1$, that is, the differential equation

$$-\frac{\mathrm{d}}{\mathrm{d}x}\left(-4\frac{\mathrm{d}u}{\mathrm{d}x}\right) + 8\frac{\mathrm{d}u}{\mathrm{d}x} + 5u = \sin x + 8\cos x, \qquad -\pi < x < \pi,$$

subject to one Neumann boundary condition

$$-4\frac{\mathrm{d}u}{\mathrm{d}x} = -20, \qquad x = -\pi$$

and one Dirichlet boundary condition

$$u = 2, \qquad x = \pi.$$

This boundary value problem has true solution

$$u(x) = 2\mathrm{e}^{-x}\left(\mathrm{e}^{\pi}\sin\frac{x}{2} + \frac{2(3-\mathrm{e}^{2\pi})}{\mathrm{e}^{\pi}}\cos\frac{x}{2}\right) + \sin x.$$

A computational implementation of the finite element method, using a linear approximation to the solution on each element, was developed in Exercise 3.1.

4.1 In this exercise, we will develop a computational implementation of the finite element method using a quadratic approximation to the solution on each element.

(a) Write down the weak formulation of the boundary value problem. State clearly the sets that the weak solution and test functions lie in.
(b) Develop a computational implementation to calculate the finite element solution of the boundary value problem that is a quadratic approximation to the solution on each element. You may like to base this implementation on Listing 4.1. Confirm that your implementation is correct by comparing the finite element solution with the true solution.
(c) Write down an expression for the L^2 error of the finite element solution. Extend your computational implementation so that it also outputs the L^2 error of the finite element solution. Use your implementation to confirm the theoretical error bound given by Theorem 4.1. If you are unsure how to do this, more details are given in Exercise 2.2 from Chap. 2, and in Sect. 3.9.

4.2 In this exercise, we will develop a computational implementation of the finite element method, using a cubic approximation to the solution on each element.

(a) Repeat Exercise 4.1 for a finite element solution that uses a cubic approximation to the solution on each element.
(b) Use your implementation with $M = 1, 2, 3, 4$ quadrature points. Does this affect the order of convergence of the solution?

4.3 An overview of the use of basis functions of a general polynomial degree to calculate a finite element solution of a differential equation was given in Sect. 4.4. Using this description, develop a finite element implementation that uses a quartic (i.e. a degree four) polynomial approximation to the solution on each element.

Chapter 5
Nonlinear Boundary Value Problems

The differential equations investigated in earlier chapters have all been linear differential equations. The finite element discretisation of these equations yielded systems of linear algebraic equations that may be solved using established, robust and reliable linear algebra techniques. In this chapter, we describe the application of the finite element method to nonlinear boundary value problems.

The key difference when applying the finite element method to nonlinear differential equations, rather than linear differential equations, is that we must solve a system of nonlinear algebraic equations. Finding the solution of a system of nonlinear algebraic equations is, in general, more challenging than for linear systems. An extensive discussion of the existence and uniqueness of the solution of a general system of nonlinear algebraic equations is beyond the scope of this book. We do, however, provide an introduction to the computation of the solution in Section A.2 of Appendix A that will be sufficient for most finite element applications.

We introduce the finite element method for nonlinear differential equations through the use of a simple example. We then present a discussion of how this implementation may be extended for more general nonlinear differential equations. Throughout this chapter, we calculate finite element solutions that are a linear approximation to the solution on each element. We leave the use of higher order basis functions, described in Chap. 4, as an exercise for the reader.

5.1 A First Example

The implementation of the finite element method for nonlinear problems may be demonstrated using the example nonlinear boundary value problem, defined by

$$-\frac{\mathrm{d}}{\mathrm{d}x}\left((1+u^2)\frac{\mathrm{d}u}{\mathrm{d}x}\right) + 2(1+x^4) + 8u^2 = 0, \qquad -1 < x < 2, \tag{5.1}$$

J. Whiteley, *Finite Element Methods*, Mathematical Engineering,
DOI 10.1007/978-3-319-49971-0_5

subject to one Dirichlet boundary condition and one natural Neumann boundary condition:

$$u = 1, \qquad x = -1, \tag{5.2}$$

$$(1+u^2)\frac{\mathrm{d}u}{\mathrm{d}x} = 68, \qquad x = 2. \tag{5.3}$$

The solution of this boundary value problem is $u = x^2$.

As in earlier chapters, we have chosen to apply one Dirichlet boundary condition and one Neumann boundary condition in our model problem. This allows us to illustrate how both categories of boundary condition are handled. The material presented here may easily be adapted to any suitable combination of these boundary conditions.

5.1.1 *The Weak Formulation*

As with linear differential equations, we begin by writing the boundary value problem given by Eqs. (5.1)–(5.3) in weak form using the Sobolev space, and subsets of the Sobolev space, defined in Sect. 3.2.1. We multiply the differential equation by a function $v(x) \in H_0^1(-1, 2)$ and integrate the resulting product over the computational domain $-1 < x < 2$. After an application of integration by parts, we obtain

$$-\left[(1+u^2)\frac{\mathrm{d}u}{\mathrm{d}x}v\right]_{-1}^{2} + \int_{-1}^{2} (1+u^2)\frac{\mathrm{d}u}{\mathrm{d}x}\frac{\mathrm{d}v}{\mathrm{d}x} + 2(1+x^4)v + 8u^2 v \,\mathrm{d}x = 0. \tag{5.4}$$

Our boundary value problem specifies one Dirichlet boundary condition, Eq. (5.2), that applies at $x = -1$. As $v(x) \in H_0^1(-1, 2)$, we have $v(x) = 0$ at any value of x where a Dirichlet boundary condition is applied, and so $v(-1) = 0$. Using this condition on $v(-1)$, and the Neumann boundary condition given by Eq. (5.3), we may write Eq. (5.4) as

$$\int_{-1}^{2} (1+u^2)\frac{\mathrm{d}u}{\mathrm{d}x}\frac{\mathrm{d}v}{\mathrm{d}x} + 2(1+x^4)v + 8u^2 v \,\mathrm{d}x - 68v(2) = 0.$$

So far we have used the differential equation, Eq. (5.1), and the Neumann boundary condition, Eq. (5.3). We have not yet used the Dirichlet boundary condition, other than to ensure that test functions $v(x)$ lie in the appropriate set of functions. We complete the weak formulation by insisting that the weak solution satisfies the Dirichlet boundary condition, given by Eq. (5.2). The weak solution of Eqs. (5.1)–(5.3) is then given by:

find $u(x) \in H_E^1(-1, 2)$ such that

$$\int_{-1}^{2} (1+u^2)\frac{\mathrm{d}u}{\mathrm{d}x}\frac{\mathrm{d}v}{\mathrm{d}x} + 2(1+x^4)v + 8u^2 v \, \mathrm{d}x - 68v(2) = 0, \tag{5.5}$$

for all functions $v(x) \in H_0^1(-1, 2)$.

5.1.2 The Mesh, Basis Functions and Finite Element Solution

We will calculate a finite element solution that is a linear approximation to the solution of the boundary value problem on each element. The mesh and basis functions described in Chap. 3 are suitable for this purpose. We therefore use the mesh of N elements and $N+1$ nodes described in Sect. 3.3.1, and the basis functions $\phi_j(x)$, $j = 1, 2, \ldots, N+1$, described in Sect. 3.3.2.2. Note that we use the definition of the global basis functions in terms of local basis functions, defined on the canonical element $0 \le X \le 1$, and given by Eqs. (3.2.1)–(3.2.6). This will aid the integration over individual elements that is required when setting up the system of algebraic equations later in this chapter.

The set S^h takes the same definition as in Sect. 3.3.3. This set contains all functions $V(x)$ that can be written in the form

$$V(x) = \sum_{j=1}^{N+1} V_j \phi_j(x),$$

where $\phi_j(x)$, $j = 1, 2, \ldots, N+1$ are defined by Eqs. (3.2.1)–(3.2.6). Further, as in Sect. 3.3.3, the set S_E^h is the subset of S^h that includes only functions that satisfy the Dirichlet boundary condition, Eq. (5.2), and the set S_0^h is the subset of S^h that includes only functions that are zero where Dirichlet boundary conditions are applied.

Having defined the sets S_E^h and S_0^h, and using the weak formulation given by Eq. (5.5), we may define the finite element solution of Eqs. (5.1)–(5.3) by:

find $U(x) \in S_E^h$ such that

$$\int_{-1}^{2} (1+U^2)\frac{\mathrm{d}U}{\mathrm{d}x}\frac{\mathrm{d}\phi_i}{\mathrm{d}x} + 2(1+x^4)\phi_i + 8U^2\phi_i \, \mathrm{d}x - 68\phi_i(2) = 0, \tag{5.6}$$

for all functions $\phi_i(x) \in S_0^h$.

5.1.3 The Algebraic Equations

We write the finite element solution as the member of S^h given by

$$U(x) = \sum_{j=1}^{N+1} U_j \phi_j(x), \tag{5.7}$$

where $\phi_j(x)$, $j = 1, 2, \ldots, N+1$, are defined by Eqs. (3.21)–(3.26). We will now assemble $N+1$ equations to allow us to determine the $N+1$ unknown values $U_1, U_2, \ldots, U_{N+1}$. As with the systems of algebraic equations that were derived from linear differential equations in earlier chapters, these equations will comprise: (i) equations that ensure that $U(x)$ satisfies the Dirichlet boundary conditions, i.e. $U(x) \in S_E^h$; and (ii) equations that arise from substituting a suitable test function $\phi_i(x)$ into Eq. (5.6).

The first condition on $U(x)$ given in Eq. (5.6) is that $U(x) \in S_E^h$. As $U(x)$ is specified by Eq. (5.7), it already lies in the set S^h. To ensure that $U(x) \in S_E^h$, we demand that $U(x)$ satisfies all Dirichlet boundary conditions. We have one Dirichlet boundary condition, given by Eq. (5.2), which is applied at the node $x_1 = -1$. Using Eq. (3.37), this boundary condition is satisfied provided that

$$U_1 = 1. \tag{5.8}$$

The basis functions $\phi_i(x)$, $i = 2, 3, \ldots, N+1$, all satisfy the condition $\phi_i(-1) = 0$ and therefore lie in the set S_0^h. These basis functions may be substituted into Eq. (5.6) to deduce that, for $i = 2, 3, \ldots, N+1$:

$$\int_{-1}^{2} (1+U^2)\frac{\mathrm{d}U}{\mathrm{d}x}\frac{\mathrm{d}\phi_i}{\mathrm{d}x} + 2(1+x^4)\phi_i + 8U^2\phi_i \,\mathrm{d}x - 68\phi_i(2) = 0. \tag{5.9}$$

Noting that $x_{N+1} = 2$, on using Eq. (3.20) we see that

$$\phi_i(2) = \begin{cases} 0, & i = 2, 3, \ldots, N, \\ 1, & i = N+1. \end{cases}$$

We may therefore write Eq. (5.9) as

$$\int_{-1}^{2} (1+U^2)\frac{\mathrm{d}U}{\mathrm{d}x}\frac{\mathrm{d}\phi_i}{\mathrm{d}x} + 2(1+x^4)\phi_i + 8U^2\phi_i \,\mathrm{d}x = 0, \quad i = 2, 3, \ldots, N, \tag{5.10}$$

and

$$\int_{-1}^{2} (1+U^2)\frac{\mathrm{d}U}{\mathrm{d}x}\frac{\mathrm{d}\phi_{N+1}}{\mathrm{d}x} + 2(1+x^4)\phi_{N+1} + 8U^2\phi_{N+1} \,\mathrm{d}x - 68 = 0. \tag{5.11}$$

Combining Eqs. (5.8), (5.10) and (5.11), we have $N+1$ algebraic equations for the $N+1$ unknown values $U_1, U_2, \ldots, U_{N+1}$, as required.

5.1.4 Assembling the Algebraic Equations

We now explain how the system of algebraic equations, given by Eqs. (5.8), (5.10) and (5.11), is assembled in practice. These equations contain the same features as those seen in earlier chapters: (i) contributions from integrals over the whole domain (in Eqs. (5.10) and (5.11)); (ii) a contribution from the Neumann boundary condition (in Eq. (5.11)) that includes only a known explicit term; and (iii) an algebraic equation to satisfy the Dirichlet boundary condition where all terms may be written down explicitly (Eq. (5.8)).

As in earlier chapters, we decompose the integrals that appear in Eqs. (5.10) and (5.11) into the sum of the contributions from each element. We write Eq. (5.10) as, for $i = 2, 3, \ldots, N$,

$$\sum_{k=1}^{N} \int_{x_k}^{x_{k+1}} (1 + U^2)\frac{\mathrm{d}U}{\mathrm{d}x}\frac{\mathrm{d}\phi_i}{\mathrm{d}x} + 2(1 + x^4)\phi_i + 8U^2\phi_i \,\mathrm{d}x = 0. \tag{5.12}$$

Contributions to the integral in Eq. (5.11) may be written in a similar manner. The contribution to this sum from element k, i.e. the integral over $x_k \leq x \leq x_{k+1}$, may be calculated by mapping this element to the canonical element $0 \leq X \leq 1$ using the change of variable

$$x = x_k + hX, \qquad 0 \leq X \leq 1, \tag{5.13}$$

where $h = x_{k+1} - x_k$. Remembering, from Eqs. (3.22) and (3.23), that the local basis functions are given by

$$\phi_{\text{local},1}(X) = 1 - X, \tag{5.14}$$

$$\phi_{\text{local},2}(X) = X, \tag{5.15}$$

the finite element solution on element k is

$$U(X) = U_k\phi_{\text{local},1}(X) + U_{k+1}\phi_{\text{local},2}(X). \tag{5.16}$$

We see from Eq. (5.12) that the term U^2 appears in the integral over each element. On element k, this quantity is given by

$$U^2 = U_k^2\phi_{\text{local},1}^2(X) + 2U_kU_{k+1}\phi_{\text{local},1}(X)\phi_{\text{local},2}(X) + U_{k+1}^2\phi_{\text{local},2}^2(X). \tag{5.17}$$

It is clear from this expression for U^2 that the integral over each element required to compute Eq. (5.12) is a nonlinear function of the values $U_1, U_2, \ldots, U_{N+1}$. As a consequence, it is not possible to write the system of algebraic equations as the linear system $A\mathbf{U} = \mathbf{b}$ as we did in Chaps. 2, 3 and 4. Instead, we write Eqs. (5.8), (5.10) and (5.11) as the nonlinear system

$$\mathbf{R}(\mathbf{U}) = \mathbf{0}, \tag{5.18}$$

where the vector **R**—commonly known as the *residual vector*—has entries given by

$$R_1(\mathbf{U}) = U_1 - 1, \tag{5.19}$$

$$R_i(\mathbf{U}) = \sum_{k=1}^{N} \int_{x_k}^{x_{k+1}} (1+U^2)\frac{\mathrm{d}U}{\mathrm{d}x}\frac{\mathrm{d}\phi_i}{\mathrm{d}x} + 2(1+x^4)\phi_i + 8U^2\phi_i \,\mathrm{d}x, \quad i = 2, 3, \ldots, N, \tag{5.20}$$

$$R_{N+1}(\mathbf{U}) = \sum_{k=1}^{N} \int_{x_k}^{x_{k+1}} (1+U^2)\frac{\mathrm{d}U}{\mathrm{d}x}\frac{\mathrm{d}\phi_{N+1}}{\mathrm{d}x} + 2(1+x^4)\phi_{N+1} + 8U^2\phi_{N+1} \,\mathrm{d}x - 68. \tag{5.21}$$

An elementary discussion of methods for solving systems of nonlinear algebraic equations, such as Eq. (5.18), is given in Appendix A.2. The easiest techniques to use when solving these equations, for example, the function `fsolve` provided by MATLAB, require only specification of the function $\mathbf{R}(\mathbf{V})$ for a given $\mathbf{V}$, and an initial guess $\mathbf{U}_0$ to the solution $\mathbf{U}$. We now explain how the function $\mathbf{R}(\mathbf{U})$ may be evaluated, allowing the nonlinear system of equations given by Eq. (5.18) to be solved using a suitable technique, such as the function `fsolve` provided by MATLAB. We assemble these equations, as in earlier chapters, by first evaluating the integrals over the whole domain that appear in Eqs. (5.20) and (5.21). We then add in the extra contribution to Eq. (5.21) that arises from the Neumann boundary condition, before setting Eq. (5.19) to satisfy the Dirichlet boundary conditions.

5.1.4.1 Evaluating the Integrals on Each Element

Equations (5.20) and (5.21) reveal that, on element k, we have to calculate a contribution to the residual of the form

$$\int_{x_k}^{x_{k+1}} (1+U^2)\frac{\mathrm{d}U}{\mathrm{d}x}\frac{\mathrm{d}\phi_i}{\mathrm{d}x} + 2(1+x^4)\phi_i + 8U^2\phi_i \,\mathrm{d}x, \tag{5.22}$$

where i takes the values $2, 3, \ldots, N+1$. As in Chaps. 2 and 3 we observe that, on element k, the only basis functions that are nonzero are ϕ_k and ϕ_{k+1}. The local contribution to the residual from element k is then nonzero only for entries generated with $i = k$ and $i = k+1$. As a consequence, we need only calculate the contribution from this element to Eq. (5.22) using $i = k$ and $i = k+1$, which we denote by $R^{(k)}_{\text{local},1}(\mathbf{U})$ and $R^{(k)}_{\text{local},2}(\mathbf{U})$, respectively. These local contributions are given by

$$R^{(k)}_{\text{local},1}(\mathbf{U}) = \int_{x_k}^{x_{k+1}} (1+U^2)\frac{\mathrm{d}U}{\mathrm{d}x}\frac{\mathrm{d}\phi_k}{\mathrm{d}x} + 2(1+x^4)\phi_k + 8U^2\phi_k \,\mathrm{d}x, \tag{5.23}$$

$$R^{(k)}_{\text{local},2}(\mathbf{U}) = \int_{x_k}^{x_{k+1}} (1+U^2)\frac{\mathrm{d}U}{\mathrm{d}x}\frac{\mathrm{d}\phi_{k+1}}{\mathrm{d}x} + 2(1+x^4)\phi_{k+1} + 8U^2\phi_{k+1} \,\mathrm{d}x. \tag{5.24}$$

The local residuals, given by Eqs. (5.23) and (5.24), may easily be evaluated by mapping element k to the canonical element $0 \le X \le 1$, using the change of variable given by Eq. (5.13). On element k, the basis functions ϕ_k and ϕ_{k+1} are given by Eqs. (5.14) and (5.15), and so the local contributions to the residual may be written

$$R^{(k)}_{\text{local},1}(\mathbf{U}) = h \int_0^1 \frac{1}{h^2}(1+U^2)\frac{\mathrm{d}U}{\mathrm{d}X}\frac{\mathrm{d}\phi_{\text{local},1}}{\mathrm{d}X} + 2(1+(x_k+hX)^4)\phi_{\text{local},1} + 8U^2\phi_{\text{local},1}\,\mathrm{d}X, \tag{5.25}$$

$$R^{(k)}_{\text{local},2}(\mathbf{U}) = h \int_0^1 \frac{1}{h^2}(1+U^2)\frac{\mathrm{d}U}{\mathrm{d}X}\frac{\mathrm{d}\phi_{\text{local},2}}{\mathrm{d}X} + 2(1+(x_k+hX)^4)\phi_{\text{local},2} + 8U^2\phi_{\text{local},2}\,\mathrm{d}X, \tag{5.26}$$

where U^2 is given by Eq. (5.17), and

$$\frac{\mathrm{d}U}{\mathrm{d}X} = U_k\frac{\mathrm{d}\phi_{\text{local},1}}{\mathrm{d}X} + U_{k+1}\frac{\mathrm{d}\phi_{\text{local},2}}{\mathrm{d}X}. \tag{5.27}$$

The integrals given by Eqs. (5.25) and (5.26) may now be evaluated, either analytically or using quadrature. We then add these local contributions into the global residual by incrementing the appropriate entry of $\mathbf{R}$. The local contribution $R^{(k)}_{\text{local},1}$ from element k was the integral generated from Eq. (5.22) with $i = k$: hence, $R^{(k)}_{\text{local},1}$ should be added to entry R_k of the global residual. Similarly, the local contribution $R^{(k)}_{\text{local},2}$ from element k was the integral generated from Eq. (5.22) with $i = k+1$ and should be added to entry R_{k+1} of the global residual.

The integral contributions to $\mathbf{R}(\mathbf{U})$ may therefore be computed by looping over all elements, calculating all nonzero local contributions as described above, and then adding these local contributions into the global residual vector.

5.1.4.2 Setting the Entries Defined Explicitly

To complete specification of $\mathbf{R}(\mathbf{U})$, we need to: (i) add the final term into Eq. (5.21) to satisfy the Neumann boundary condition; and (ii) specify R_1, given by Eq. (5.19), to satisfy the Dirichlet boundary conditions. We now give an example computational implementation of the material presented in this section.

5.1.5 Computational Implementation

In Listing 5.1, we provide a computational implementation of the material discussed in this chapter for calculating the finite element solution of Eqs. (5.1)–(5.3). As with

MATLAB implementations in earlier chapters, the variable names correspond to the notation used when presenting the material.

The listing begins by requiring the user to provide an appropriate mesh **x**, and initial guess to the solution at each node in the mesh $\mathbf{U_0}$. The vector **x** that specifies the mesh should satisfy the properties listed in Sect. 3.3.1: (i) if there are N elements in the mesh, then this vector should be of length $N+1$; (ii) $x_1 = -1$; (iii) $x_{N+1} = 2$; and (iv) the entries of the vector should be listed in ascending order. The in-built MATLAB function `fsolve` is then called on line 5. This function finds the finite element solution **U**, starting from the initial guess $\mathbf{U_0}$, such that all components of the vector-valued function `CalculateResidual` are zero, as is required by Eq. (5.18)—see the MATLAB documentation for more details. The solution **U** is returned as the output of the MATLAB function.

The function `CalculateResidual` begins at line 15 of the listing. The function first deduces the number of elements (line 18) and initialises the variable `Residual` to be a vector of the correct size, with all entries set to zero (line 21). We then loop over all elements in lines 25–28, calculating the nonzero local contributions to the residual from each element, before incrementing the appropriate entries of the global residual with these local contributions. Finally, this function adds the contribution to entry $N+1$ of the global vector that is required to satisfy the Neumann boundary condition at $x = 2$ (line 31) and sets the first entry of this vector to satisfy the Dirichlet boundary condition at $x = -1$ (line 34).

The local contributions to the residual on each element are calculated using the function `CalculateResidualOnElement` that starts on line 39 of the listing. We begin this function by calculating the element length (line 42), and defining the numerical quadrature scheme used (lines 45–47), before initialising the local contribution to the residual to zero (line 50). We then loop over the quadrature points, calculating the contributions to the local residuals (given by Eqs. (5.25) and (5.26)) from each of these points—we first evaluate the quantities that appear in these equations (lines 60–65), before adding suitably weighted function evaluations into the local contributions to the residual (lines 68–75).

Having calculated **U**, we then plot the solution (lines 8–10).

Suppose we want to calculate the finite element solution of Eqs. (5.1)–(5.3), using a uniform mesh with 100 elements, and an initial guess to the solution that is zero at each node. The MATLAB implementation given in Listing 5.1 may be called by typing

```
x = linspace(-1,2,101);
U0 = zeros(101,1);
U = Chap5_CalculateFemForNonlinearBVP(x, U0);
```

into a MATLAB session.

Note that we did not put much effort into specifying our initial guess to the solution, simply setting this guess to be zero at each node. In this case, the MATLAB routine used to solve the system of nonlinear equations found the solution from this initial guess. Readers should be aware, however, that for some problems a more accurate initial guess to the solution may be required.

Listing 5.1 `Chap5_CalculateFemForNonlinearBVP.m`, a MATLAB function for calculating the finite element solution of Eqs. (5.1)–(5.3), specifying only the residual of the nonlinear system.

```
% Specify mesh x and initial guess U0
function U = Chap5_CalculateFemForNonlinearBVP(x, U0)

% Use Matlab routine to solve nonlinear algebraic system
U = fsolve(@CalculateResidual, U0, [], x);

% Plot solution
plot(x, U, '-')
xlabel('x')
ylabel('U')



% Function for calculating global residual
function Residual = CalculateResidual(U, x)

% Deduce number of elements
N = length(x) - 1;

% Initialise residual to zero
Residual = zeros(N+1, 1);

% Loop over elements, computing local contribution to
% residual and incrementing global residual
for k=1:N
    LocalResidual = CalculateResidualOnElement(U, x, k);
    Residual(k:k+1) = Residual(k:k+1) + LocalResidual;
end

% Handle Neumann boundary condition at x=2
Residual(N+1) = Residual(N+1) - 68;

% Handle Dirichlet boundary condition at x=-1
Residual(1) = U(1) - 1;



% Function for calculating local residual
function LocalResidual = CalculateResidualOnElement(U, x, k)

% Set element length h
h = x(k+1) - x(k);

% Define quadrature rule
M = 3;
QuadWeights = [5/18, 4/9, 5/18];
QuadPoints = [0.5*(1-sqrt(0.6)), 0.5, 0.5*(1+sqrt(0.6))];

% Initialise local residual to zero
LocalResidual = zeros(2,1);

% Loop over quadrature points to evaluate integration
% using Gaussian quadrature
for m=1:M
   % Set local coordinate X
   X = QuadPoints(m);
```

```

    % Calculate local values of basis functions,
    % their derivatives, x, U and the derivative of U
    phi_local = [1-X, X];
    phi_prime_local = [-1, 1];
    x_local = x(k) + h*X;
    LocalU = U(k)*phi_local(1) + U(k+1)*phi_local(2);
    LocalUPrime = U(k)*phi_prime_local(1) + ...
        U(k+1)*phi_prime_local(2);

    % Evaluate integrand and add into sum
    for i=1:2
        LocalResidual(i) = LocalResidual(i) + ...
            h*QuadWeights(m)* ...
            (1/h/h*(1+LocalU*LocalU)* ...
            LocalUPrime*phi_prime_local(i) + ...
            2*(1+x_local^4)*phi_local(i) + ...
            8*LocalU*LocalU*phi_local(i));
    end
end
```

5.1.6 Calculation of the Jacobian Matrix

In Sect. 5.1.5, we provided a computational implementation for calculating the finite element solution of Eqs. (5.1)–(5.3). We used the MATLAB function `fsolve` to solve the system of nonlinear equations given by Eq. (5.18), using specification only of the residual vector $\mathbf{R}$. Some numerical methods for solving nonlinear systems of algebraic equations require specification not just of the residual vector, but also the *Jacobian matrix*, denoted by J, with entries given by

$$J_{i,j}(\mathbf{U}) = \frac{\partial R_i}{\partial U_j}, \qquad i, j = 1, 2, \ldots, N+1. \tag{5.28}$$

Even if the numerical method being used does not require the Jacobian matrix to be specified, it may provide the option to specify this matrix; the MATLAB function `fsolve` (that was used in Listing 5.1) allows the user to do this. Supplying the Jacobian matrix will usually improve the efficiency and the reliability of the root-finding algorithm.

We now explain how entries of the Jacobian matrix, defined by Eq. (5.28), may be calculated. Clearly from Eq. (5.19), the first row of the Jacobian matrix is given by, for $j = 1, 2, \ldots, N+1$,

$$J_{1,j} = \begin{cases} 1, & j = 1, \\ 0, & j \neq 1. \end{cases} \tag{5.29}$$

We now consider rows $2, 3, \ldots, N+1$ of the Jacobian matrix. Using Eqs. (5.20) and (5.21), we see that, for $i = 2, 3, \ldots, N+1$, the entries of the Jacobian matrix can be expressed as the sum of the contributions from each element:

$$J_{i,j} = \sum_{k=1}^{N} \frac{\partial}{\partial U_j} \int_{x_k}^{x_{k+1}} (1+U^2)\frac{\mathrm{d}U}{\mathrm{d}x}\frac{\mathrm{d}\phi_i}{\mathrm{d}x} + 2(1+x^4)\phi_i + 8U^2\phi_i \,\mathrm{d}x, \qquad j - 1, 2, \ldots, N+1, \tag{5.30}$$

where we have interchanged the order of summation and partial differentiation.

As the reader would expect, the entries of Eq. (5.30) are assembled by calculating the contribution from individual elements. The contribution to $J_{i,j}$, where $i = 2, 3, \ldots, N+1$, and $j = 1, 2, \ldots, N+1$, from integrating over element k is given by

$$\frac{\partial}{\partial U_j} \int_{x_k}^{x_{k+1}} (1+U^2)\frac{\mathrm{d}U}{\mathrm{d}x}\frac{\mathrm{d}\phi_i}{\mathrm{d}x} + 2(1+x^4)\phi_i + 8U^2\phi_i \,\mathrm{d}x. \tag{5.31}$$

Remembering that $\phi_k(x)$ and $\phi_{k+1}(x)$ are the only basis functions that are nonzero on element k, the integrand above will be zero unless $i = k$ or $i = k+1$. We therefore need only calculate the contributions to this integral using $i = k$ and $i = k+1$, as all other entries will be zero. Recalling the definition of the local contributions to the residual on element k, denoted by $R^{(k)}_{\text{local},1}(\mathbf{U})$, $R^{(k)}_{\text{local},2}(\mathbf{U})$ and defined by Eqs. (5.23) and (5.24), we see that the nonzero contributions to Eq. (5.31) when $i = k$ are, for $j = 1, 2, \ldots, N+1$,

$$\frac{\partial R^{(k)}_{\text{local},1}}{\partial U_j}, \tag{5.32}$$

and, when $i = k+1$, the contributions to Eq. (5.31) are, for $j = 1, 2, \ldots, N+1$,

$$\frac{\partial R^{(k)}_{\text{local},2}}{\partial U_j}. \tag{5.33}$$

Further, on element k, the finite element solution $U(x)$ may be written

$$U(x) = U_k\phi_k(x) + U_{k+1}\phi_{k+1}(x), \tag{5.34}$$

and so the local contributions from element k to the Jacobian given by Eq. (5.30) depend only on the entries U_k and U_{k+1} of the vector $\mathbf{U}$. As a consequence, the expressions given by Eqs. (5.32) and (5.33) are zero unless $j = k$ or $j = k+1$. We now see that the local contribution from element k to the global Jacobian matrix, given by Eq. (5.30), is nonzero for only four entries of J: $J_{k,k}$, $J_{k,k+1}$, $J_{k+1,k}$ and

$J_{k+1,k+1}$. As such, only these entries need be calculated. They may be stored in the 2×2 matrix denoted by $J_{\text{local}}^{(k)}$, with entries given by

$$J_{\text{local}}^{(k)} = \begin{pmatrix} \frac{\partial R_{\text{local},1}^{(k)}}{\partial U_k} & \frac{\partial R_{\text{local},1}^{(k)}}{\partial U_{k+1}} \\ \frac{\partial R_{\text{local},2}^{(k)}}{\partial U_k} & \frac{\partial R_{\text{local},2}^{(k)}}{\partial U_{k+1}} \end{pmatrix}, \tag{5.35}$$

where $J_{\text{local},1,1}^{(k)}$ contributes to $J_{k,k}$, $J_{\text{local},1,2}^{(k)}$ contributes to $J_{k,k+1}$, $J_{\text{local},2,1}^{(k)}$ contributes to $J_{k+1,k}$, and $J_{\text{local},2,2}^{(k)}$ contributes to $J_{k+1,k+1}$.

The partial derivatives that appear in Eq. (5.35) are straightforward to evaluate by partial differentiation of Eqs. (5.25) and (5.26). Using Eqs. (5.27) and (5.34), we see that when element k is mapped to the canonical interval $0 \leq X \leq 1$

$$\frac{\partial U}{\partial U_k} = \phi_{\text{local},1}(X), \qquad \frac{\partial U}{\partial U_{k+1}} = \phi_{\text{local},2}(X),$$

$$\frac{\partial}{\partial U_k}\left(\frac{\mathrm{d}U}{\mathrm{d}X}\right) = \frac{\mathrm{d}\phi_{\text{local},1}}{\mathrm{d}X}, \qquad \frac{\partial}{\partial U_{k+1}}\left(\frac{\mathrm{d}U}{\mathrm{d}X}\right) = \frac{\mathrm{d}\phi_{\text{local},2}}{\mathrm{d}X}.$$

Hence, using the chain rule,

$$\begin{aligned} J_{\text{local},1,1}^{(k)} &= \frac{\partial R_{\text{local},1}^{(k)}}{\partial U_k} \\ &= h\frac{\partial}{\partial U_k}\int_0^1 \frac{1}{h^2}(1+U^2)\frac{\mathrm{d}U}{\mathrm{d}X}\frac{\mathrm{d}\phi_{\text{local},1}}{\mathrm{d}X} + 2(1+(x_k+hX)^4)\phi_{\text{local},1} + \\ &\quad 8U^2\phi_{\text{local},1}\,\mathrm{d}X, \\ &= h\int_0^1 \frac{1}{h^2}\left(\left(2U\frac{\partial U}{\partial U_k}\right)\frac{\mathrm{d}U}{\mathrm{d}X} + (1+U^2)\frac{\partial}{\partial U_k}\left(\frac{\mathrm{d}U}{\mathrm{d}X}\right)\right)\frac{\mathrm{d}\phi_{\text{local},1}}{\mathrm{d}X} + \\ &\quad 16U\frac{\partial U}{\partial U_k}\phi_{\text{local},1}\,\mathrm{d}X \\ &= h\int_0^1 \frac{1}{h^2}\left((2U\phi_{\text{local},1})\frac{\mathrm{d}U}{\mathrm{d}X} + (1+U^2)\frac{\mathrm{d}\phi_{\text{local},1}}{\mathrm{d}X}\right)\frac{\mathrm{d}\phi_{\text{local},1}}{\mathrm{d}X} + \\ &\quad 16U\phi_{\text{local},1}\phi_{\text{local},1}\,\mathrm{d}X. \end{aligned}$$

More generally, for $i = 1, 2$, and $j = 1, 2$:

$$\begin{aligned} J_{\text{local},i,j}^{(k)} = h\int_0^1 \frac{1}{h^2}\left((2U\phi_{\text{local},j})\frac{\mathrm{d}U}{\mathrm{d}X} + (1+U^2)\frac{\mathrm{d}\phi_{\text{local},j}}{\mathrm{d}X}\right)\frac{\mathrm{d}\phi_{\text{local},i}}{\mathrm{d}X} + \\ 16U\phi_{\text{local},j}\phi_{\text{local},i}\,\mathrm{d}X. \end{aligned} \tag{5.36}$$

These integrals may easily be evaluated using quadrature.

The global Jacobian may therefore be computed by looping over all elements, calculating the local nonzero contributions on each element, before adding these local contributions into the global matrix J. Having done this, we set the entries corresponding to rows of the residual that enforce Dirichlet boundary conditions, that is, the entries given by Eq. (5.29).

5.1.6.1 Computational Implementation

In Listing 5.1, we presented a computational implementation for calculating the finite element solution of Eqs. (5.1)–(5.3), solving the resulting nonlinear system of algebraic equations given by Eq. (5.18) by specifying only the residual vector. In Listing 5.2, we extend this computational implementation, solving the nonlinear system of algebraic equations by specification of both the residual vector and Jacobian matrix. We assume that the reader is familiar with the implementation given in Listing 5.1 and describe only the modifications made in the listing below.

The mesh and initial guess to the solution should be set as described in Sect. 5.1.5. Specification of only the residual vector is the default for the MATLAB function `fsolve`, and so we set the option that allows `fsolve` to use the Jacobian matrix in line 8, and pass this option to `fsolve` in line 10.

We write a function `CalculateResidualAndJacobian` in lines 20–46 of the listing that computes both the global residual vector and global Jacobian matrix. This function initialises both of these to zero (lines 27–28), before looping over all elements in lines 32–38, calculating the nonzero local contributions and adding these contributions into the correct locations in the global residual vector and global Jacobian matrix. Finally, the boundary conditions are handled by lines 41–46.

The function `CalculateResidualAndJacobianOnElement` (lines 52–102) calculates the local contribution to both the residual vector and Jacobian matrix on each element, using Gaussian quadrature. This function follows the pattern of the function `CalculateResidualOnElement` from Listing 5.1, but with the addition of calculating the entries of the local Jacobian matrix given by Eq. (5.36) in lines 91–99 of the listing by using Gaussian quadrature.

This MATLAB function may be called by typing, as in Sect. 5.1.5,

```
x = linspace(-1,2,101);
U0 = zeros(101,1);
U = Chap5_CalculateFemForNonlinearBVP_Jacobian(x, U0);
```

into a MATLAB session. We encourage the reader to execute both Listings 5.1 and 5.2 for meshes containing different numbers of elements, and to record the execution time for both functions. The reader will observe that specification of the Jacobian matrix allows the finite element solution to be computed much more efficiently—see Exercise 5.1.

Listing 5.2 `Chap5_CalculateFemForNonlinearBVP_Jacobian.m`, a MATLAB function for calculating the finite element solution of Eqs. (5.1)–(5.3), specifying both the residual vector and Jacobian matrix of the nonlinear system.

```
% Specify mesh x and initial guess U0
function U = ...
    Chap5_CalculateFemForNonlinearBVP_Jacobian(x, U0)

% Use Matlab routine to solve nonlinear algebraic system

% Specify that the Jacobian will be specified by the user
options = optimoptions('fsolve','Jacobian','on');

U = fsolve(@CalculateResidualAndJacobian, U0, options, x);

% Plot solution
plot(x, U, '-')
xlabel('x')
ylabel('U')



% Function for calculating global residual and Jacobian
function [Residual, Jacobian] = ...
    CalculateResidualAndJacobian(U, x)

% Deduce number of elements
N = length(x) - 1;

% Initialise residual and Jacobian to zero
Residual = zeros(N+1, 1);
Jacobian = zeros(N+1, N+1);

% Loop over elements, computing local contribution to
% residual and incrementing global residual and Jacobian
for k=1:N
    [LocalResidual, LocalJacobian] = ...
        CalculateResidualAndJacobianOnElement(U, x, k);
    Residual(k:k+1) = Residual(k:k+1) + LocalResidual;
    Jacobian(k:k+1, k:k+1) = Jacobian(k:k+1, k:k+1) + ...
        LocalJacobian;
end

% Handle Neumann boundary condition at x=2
Residual(N+1) = Residual(N+1) - 68;

% Handle Dirichlet boundary condition at x=-1
Residual(1) = U(1) - 1;
Jacobian(1,:) = 0;
Jacobian(1,1) = 1;



% Function for calculating local contribution to
% residual and Jacobian
function [LocalResidual, LocalJacobian] = ...
    CalculateResidualAndJacobianOnElement(U, x, k)

% Set element length h
h = x(k+1) - x(k);
```

```

% Define quadrature rule
M = 3;
QuadWeights = [5/18, 4/9, 5/18];
QuadPoints = [0.5*(1-sqrt(0.6)), 0.5, 0.5*(1+sqrt(0.6))];

% Initialise local residual and Jacobian to zero
LocalResidual = zeros(2,1);
LocalJacobian = zeros(2,2);

% Loop over quadrature points to evaluate integral
% using Gaussian quadrature
for m=1:M
   % Set local coordinate X
   X = QuadPoints(m);

   % Calculate local values of basis functions,
   % their derivatives, x, U and the derivative of U
   phi_local = [1-X, X];
   phi_prime_local = [-1, 1];
   x_local = x(k) + h*X;
   LocalU = U(k)*phi_local(1) + U(k+1)*phi_local(2);
   LocalUPrime = U(k)*phi_prime_local(1) + ...
      U(k+1)*phi_prime_local(2);

   % Evaluate integrand and add into sum
   for i=1:2
       LocalResidual(i) = LocalResidual(i) + ...
          h*QuadWeights(m)* ...
          (1/h/h*(1+LocalU*LocalU)* ...
          LocalUPrime*phi_prime_local(i) + ...
          2*(1+x_local^4)*phi_local(i) + ...
          8*LocalU*LocalU*phi_local(i));

      for j=1:2
           LocalJacobian(i,j) = LocalJacobian(i,j) + ...
               h*QuadWeights(m)* ...
               (1/h/h*(2*LocalU*phi_local(j)* ...
               LocalUPrime + ...
               (1+LocalU*LocalU)*phi_prime_local(j))* ...
               phi_prime_local(i) + ...
               16*LocalU*phi_local(j)*phi_local(i));
      end

   end
end
```

5.1.7 Numerical Approximation of the Jacobian Matrix

In Sect. 5.1.6.1, we claimed that a nonlinear system of algebraic equations may be solved far more efficiently by specifying, in addition to the residual vector, the Jacobian matrix. This claim may be verified by completing Exercise 5.1 at the end of this chapter. Calculating the Jacobian matrix does, however, require extra programming to be undertaken. The example boundary value problem used in this chapter, defined by Eqs. (5.1)–(5.3), contains only terms that allow local contributions to the Jacobian matrix, given by Eq. (5.36), to be explicitly calculated fairly easily. For more complex differential equations, it may be very difficult to evaluate these contributions. Even if the contributions are calculated, some verification of the calculation may be useful. In both cases, a numerical approximation to the partial derivatives given by Eq. (5.28), calculated using the finite difference method, may be useful. We now explain how these partial derivatives may be estimated efficiently.

In Listing 5.1—the implementation that does not specify the Jacobian matrix—we wrote a function `CalculateResidual` that calculated the global residual. It may be tempting to calculate the global Jacobian matrix by using this function to estimate the partial derivatives needed, by using a finite difference approximation. To calculate the required partial derivatives with respect to U_j, we could use this function to calculate both $\mathbf{R}(\mathbf{U})$ and $\mathbf{R}(\mathbf{U} + \varepsilon \mathbf{e}_j)$, where $\mathbf{e}_j$ is a vector of the same size as $\mathbf{U}$, with entry $e_j = 1$, and all other entries taking the value zero. The partial derivatives with respect to U_j, i.e. the entries of column j of the Jacobian matrix, may then approximated by

$$J_{i,j} \approx \frac{R_i(\mathbf{U} + \varepsilon \mathbf{e}_j) - R_i(\mathbf{U})}{\varepsilon},$$

where ε takes a sufficiently small value for the approximation to the derivative to be accurate. Performing this for all columns j will indeed generate a numerical approximation to the Jacobian matrix. This approach is, however, very inefficient. To evaluate the partial derivatives in this way, we would first need to compute the original residual. We would then need to compute $N + 1$ perturbed residuals to approximate the partial derivatives with respect to each U_j, $j = 1, 2, \ldots, N + 1$, as demonstrated above. This would require the function `CalculateResidual` to be evaluated a total of $N + 2$ times.

A more efficient approach to approximating the Jacobian matrix is to calculate an approximation to the partial derivatives that appear in the local contribution to the Jacobian matrix from an individual element, given by Eq. (5.35). These partial derivatives may easily be approximated by the finite difference method, using the function `CalculateResidualOnElement` given in Listing 5.1. Taking this approach, we may assemble the Jacobian matrix as described in Sect. 5.1.6, using a numerical approximation to Eq. (5.35). We then supply both the residual vector and Jacobian matrix to the MATLAB function `fsolve` as in Listing 5.2.

5.2 A General Nonlinear Boundary Value Problem

We now discuss how the material that has been used to calculate the finite element solution of the example nonlinear differential equation, given by Eqs. (5.1)–(5.3), may be applied to more general nonlinear boundary value problems. The most general second-order, nonlinear, differential equation satisfied by a function $u(x)$ on the interval $a < x < b$ is of the form

$$f\left(x, u, \frac{\mathrm{d}u}{\mathrm{d}x}, \frac{\mathrm{d}^2u}{\mathrm{d}x^2}\right) = 0, \qquad a < x < b,$$

for some given function $f\left(x, u, \frac{\mathrm{d}u}{\mathrm{d}x}, \frac{\mathrm{d}^2u}{\mathrm{d}x^2}\right)$, combined with suitable boundary conditions. Throughout this book, the first step when calculating the finite element solution of a differential equation has always been to write the differential equation in weak form. To do this, we have multiplied the differential equation by a suitable test function and used integration by parts to eliminate the dependence on the second derivative of $u(x)$. Writing the equation in the general form above does not always allow us to integrate by parts. Hence, we restrict ourselves to differential equations that may be written in the form

$$-\frac{\mathrm{d}}{\mathrm{d}x}\left(p\left(x, u, \frac{\mathrm{d}u}{\mathrm{d}x}\right)\right) + r\left(x, u, \frac{\mathrm{d}u}{\mathrm{d}x}\right) = 0, \qquad a < x < b, \tag{5.37}$$

for specified functions $p\left(x, u, \frac{\mathrm{d}u}{\mathrm{d}x}\right), r\left(x, u, \frac{\mathrm{d}u}{\mathrm{d}x}\right)$. Although this is not the most general possible differential equation, most practical applications can be written in this form. We require suitable boundary conditions to completely specify the problem. As with the example given earlier in this chapter, we shall impose one Dirichlet boundary condition and one natural Neumann boundary condition:

$$u = u_a, \qquad x = a, \tag{5.38}$$

$$p\left(x, u, \frac{\mathrm{d}u}{\mathrm{d}x}\right) = p_b, \qquad x = b. \tag{5.39}$$

We now give an overview of how the finite element solution of Eq. (5.37), subject to the boundary conditions given by Eqs. (5.38) and (5.39), may be calculated.

5.2.1 *The Weak Formulation*

The weak formulation is obtained in the usual way: we multiply the differential equation, given by Eq. (5.37), by a function $v(x) \in H_0^1(a, b)$, and integrate this product over the interval $a < x < b$:

$$\int_a^b \left(-\frac{\mathrm{d}}{\mathrm{d}x} \left(p \left(x, u, \frac{\mathrm{d}u}{\mathrm{d}x} \right) \right) + r \left(x, u, \frac{\mathrm{d}u}{\mathrm{d}x} \right) \right) v(x) \,\mathrm{d}x = 0.$$

Integrating by parts then gives

$$-\left[p \left(x, u, \frac{\mathrm{d}u}{\mathrm{d}x} \right) v(x) \right]_a^b + \int_a^b p \left(x, u, \frac{\mathrm{d}u}{\mathrm{d}x} \right) \frac{\mathrm{d}v}{\mathrm{d}x} + r \left(x, u, \frac{\mathrm{d}u}{\mathrm{d}x} \right) v(x) \,\mathrm{d}x = 0. \tag{5.40}$$

As $v(x) \in H_0^1(a, b)$, we have $v(a) = 0$. The Neumann boundary condition given by Eq. (5.39), together with $v(a) = 0$, allows us to write Eq. (5.40) as

$$\int_a^b p \left(x, u, \frac{\mathrm{d}u}{\mathrm{d}x} \right) \frac{\mathrm{d}v}{\mathrm{d}x} + r \left(x, u, \frac{\mathrm{d}u}{\mathrm{d}x} \right) v(x) \,\mathrm{d}x - p_b v(b) = 0. \tag{5.41}$$

The weak solution of Eqs. (5.37)–(5.39) is then defined by:

find $u(x) \in H_E^1(a, b)$ such that

$$\int_a^b p \left(x, u, \frac{\mathrm{d}u}{\mathrm{d}x} \right) \frac{\mathrm{d}v}{\mathrm{d}x} + r \left(x, u, \frac{\mathrm{d}u}{\mathrm{d}x} \right) v(x) \,\mathrm{d}x - p_b v(b) = 0, \tag{5.42}$$

for all $v(x) \in H_0^1(a, b)$.

5.2.2 *The Nonlinear System of Algebraic Equations*

We use the same mesh, basis functions and sets S^h, S_E^h, S_0^h that were used for the example problem that we considered earlier in this chapter. The finite element solution, $U(x)$, may then be written as

$$U(x) = \sum_{j=1}^{N+1} U_j \phi_j(x).$$

The coefficients of the basis functions in the expression above, $U_1, U_2, \ldots, N_{N+1}$, are then determined by specifying the finite element solution by:

find $U(x) \in S_E^h$ such that

$$\int_a^b p \left(x, U, \frac{\mathrm{d}U}{\mathrm{d}x} \right) \frac{\mathrm{d}\phi_i}{\mathrm{d}x} + r \left(x, U, \frac{\mathrm{d}U}{\mathrm{d}x} \right) \phi_i(x) \,\mathrm{d}x - p_b \phi_i(b) = 0, \tag{5.43}$$

for all $\phi_i(x) \in S_0^h$.

As $U(x) \in S_E^h$, the finite element solution must satisfy any Dirichlet boundary conditions that have been specified. We have one Dirichlet boundary condition, specified at the node where $x = x_1$, and given by Eq. (5.38). Remembering that $U(x_1) = U_1$, we may deduce our first algebraic equation:

$$U_1 = u_a. \tag{5.44}$$

The functions $\phi_i(x)$, $i = 2, 3, \ldots, N + 1$, all satisfy $\phi_i(a) = 0$ and are therefore members of the set S_0^h. We may then substitute these basis functions into Eq. (5.43) to deduce that

$$\int_a^b p\left(x, U, \frac{\mathrm{d}U}{\mathrm{d}x}\right)\frac{\mathrm{d}\phi_i}{\mathrm{d}x} + r\left(x, U, \frac{\mathrm{d}U}{\mathrm{d}x}\right)\phi_i(x) - p_b\phi_i(b)\,\mathrm{d}x = 0, \qquad i = 2, 3, \ldots, N + 1. \tag{5.45}$$

The final term on the left-hand side of Eq. (5.45) may be simplified by remembering that

$$\begin{aligned}\phi_i(b) &= \phi_i(x_{N+1}) \\ &= \begin{cases} 1, & i = N + 1, \\ 0, & i \neq N + 1. \end{cases}\end{aligned}$$

The algebraic equations given by Eqs. (5.44) and (5.45) may now be written in the form

$$\mathbf{R}(\mathbf{U}) = \mathbf{0},$$

where the entries of $\mathbf{R}$ are given by

$$\begin{aligned}R_1 &= U_1 - u_a, \\ R_i &= \sum_{k=1}^{N}\int_{x_k}^{x_{k+1}} p\left(x, U, \frac{\mathrm{d}U}{\mathrm{d}x}\right)\frac{\mathrm{d}\phi_i}{\mathrm{d}x} + r\left(x, u, \frac{\mathrm{d}u}{\mathrm{d}x}\right)\phi_i(x)\,\mathrm{d}x, \quad i = 2, 3, \ldots, N, \\ R_{N+1} &= \sum_{k=1}^{N}\int_{x_k}^{x_{k+1}} p\left(x, U, \frac{\mathrm{d}U}{\mathrm{d}x}\right)\frac{\mathrm{d}\phi_{N+1}}{\mathrm{d}x} + r\left(x, u, \frac{\mathrm{d}u}{\mathrm{d}x}\right)\phi_{N+1}(x)\,\mathrm{d}x - p_b.\end{aligned}$$

These residuals may then be calculated, as described for the example differential equation in Sect. 5.1.4. The Jacobian matrix may, if required, be calculated as described in Sect. 5.1.6.

5.3 Exercises

5.1 Listing 5.1 contains a computational implementation for calculating the finite element solution of the boundary value problem given by Eqs. (5.1)–(5.3), specifying only the residual vector given by Eq. (5.18). Listing 5.2 also solves this boundary value problem, specifying both the residual vector and the Jacobian matrix given by Eq. (5.28). Execute the functions given by both Listings 5.1 and 5.2 for different numbers of elements: try, for example, $N = 50, 100, 150, 200, \ldots$. Record the execution time for both functions for all values of N that you use. How much more efficient is it to specify the global Jacobian matrix in addition to the global residual vector?

5.2 Using Listings 5.1 and 5.2, write a computational implementation that calculates the finite element solution of the boundary value problem given by Eqs. (5.1)–(5.3), using a numerical approximation to the Jacobian matrix as described in Sect. 5.1.7.

5.3 In this exercise, we will work with the differential equation

$$-\frac{\mathrm{d}}{\mathrm{d}x}\left(u\frac{\mathrm{d}u}{\mathrm{d}x}\right) + 2u^2 = 0, \qquad 0 < x < 1,$$

subject to the Dirichlet boundary conditions

$$\begin{aligned} u &= 1, && \text{at } x = 0, \\ u &= \mathrm{e}, && \text{at } x = 1. \end{aligned}$$

This equation has solution $u(x) = \mathrm{e}^x$.

a. Write down the weak formulation of the given differential equation. State clearly the definition of the sets $H_E^1(0, 1)$ and $H_0^1(0, 1)$ that you will require.
b. Calculate a finite element solution to the given differential equation, using a linear approximation to the solution on each element. You may want to start by using the function MATLAB `fsolve`, specifying only the residual vector as in Listing 5.1. When you are confident that this implementation is working correctly, develop your code so that it also specifies the Jacobian matrix, either calculated analytically or approximated numerically.
c. Repeat part (b) using a quadratic approximation to the solution on each element.

5.4 In this exercise, we will work with the differential equation

$$-\frac{\mathrm{d}}{\mathrm{d}x}\left(\mathrm{e}^u\left(\frac{\mathrm{d}u}{\mathrm{d}x}\right)^2\right) + (1-u^2)\mathrm{e}^u\cos x - 2\mathrm{e}^u u\frac{\mathrm{d}u}{\mathrm{d}x} = 0, \qquad 0 < x < \frac{\pi}{2},$$

subject to a Neumann boundary condition at $x = 0$, and a Dirichlet boundary condition at $x = \frac{\pi}{2}$:

$$e^u \left(\frac{du}{dx}\right)^2 = 1, \qquad \text{at } x = 0,$$

$$u = 1, \qquad \text{at } x = \frac{\pi}{2}.$$

This equation has solution $u(x) = \sin x$.

Repeat the subparts of Exercise 5.3 for this differential equation.

Chapter 6
Systems of Ordinary Differential Equations

So far we have only considered scalar boundary value problems, that is, a single differential equation that contains one unknown function $u(x)$. We now generalise the finite element method so that it may be applied to coupled systems of differential equations. We illustrate the basic principles using a simple example, before explaining how these ideas may be generalised for application to more complex systems.

6.1 A Model Problem

A model system of differential equations is, for $0 < x < \pi$,

$$-\frac{\mathrm{d}^2u_1}{\mathrm{d}x^2} - 3\frac{\mathrm{d}^2u_2}{\mathrm{d}x^2} - u_1 + 8u_2 = 20\sin 2x, \tag{6.1}$$

$$-\frac{\mathrm{d}^2u_2}{\mathrm{d}x^2} + 2u_1 - 4u_2 = 2\sin x. \tag{6.2}$$

As with scalar equations, we require suitable boundary conditions to be specified. In general, we expect one boundary condition for each equation at $x = 0$, and one boundary condition for each equation at $x = \pi$. As with scalar differential equations, these boundary conditions may be any combination of Dirichlet, Neumann and Robin boundary conditions, subject to constraints on the existence and uniqueness of the solution. The boundary conditions that we use to illustrate the application of the finite element method to the model system of equations above are, at $x = 0$:

$$-\frac{\mathrm{d}u_1}{\mathrm{d}x} - 3\frac{\mathrm{d}u_2}{\mathrm{d}x} = -7, \tag{6.3}$$

$$u_2 = 0, \tag{6.4}$$

J. Whiteley, *Finite Element Methods*, Mathematical Engineering,
DOI 10.1007/978-3-319-49971-0_6

and at $x = \pi$, we impose

$$u_1 = 0, \tag{6.5}$$

$$-\frac{\mathrm{d}u_2}{\mathrm{d}x} = -2. \tag{6.6}$$

We shall see later that the Neumann boundary condition given by Eq. (6.3) is a natural boundary condition for Eq. (6.1), and the Neumann boundary condition given by Eq. (6.6) is a natural boundary condition for Eq. (6.2). The solution of this model problem is

$$u_1(x) = \sin x, \qquad u_2(x) = \sin 2x.$$

6.2 The Weak Formulation

The method for deriving the weak formulation in earlier chapters has been to multiply the differential equation by a suitable test function and to integrate this product over the region on which the differential equation is defined. This is easily generalised for systems of equations such as Eqs. (6.1) and (6.2). We multiply the first differential equation, Eq. (6.1), by a suitable test function $v_1(x)$, and the second differential equation, Eq. (6.2), by a suitable test function $v_2(x)$. We must, however, choose suitable test functions.

We have already claimed that Eq. (6.3) is a natural Neumann boundary condition at $x = 0$ for the first differential equation, Eq. (6.1). Further, Eq. (6.5) may be used as a Dirichlet boundary condition for Eq. (6.1) at $x = \pi$. We then view Eq. (6.1), subject to the boundary conditions given by Eqs. (6.3) and (6.5), as a boundary value problem that satisfies a Neumann boundary condition at $x = 0$, and a Dirichlet boundary condition at $x = \pi$. We therefore specify v_1 to be a member of the set $H^1(0, \pi)$ that satisfies the condition

$$v_1(\pi) = 0, \tag{6.7}$$

so that $v_1(x) = 0$ at values of x where Dirichlet boundary conditions are applied to the first differential equation, given by Eq. (6.1).

Similarly, we will solve the second differential equation, Eq. (6.2), subject to the Dirichlet boundary condition given by Eq. (6.4) at $x = 0$, and the Neumann boundary condition given by Eq. (6.6) at $x = \pi$. Using these boundary conditions, we see that an appropriate test function $v_2(x)$ belongs to the set $H^1(0, \pi)$ and also satisfies

$$v_2(0) = 0. \tag{6.8}$$

We now multiply Eq. (6.1) by the test function $v_1(x)$ defined above and integrate the resulting product over $0 < x < \pi$. After integrating by parts, we obtain

$$\left[-\left(\frac{\mathrm{d}u_1}{\mathrm{d}x}+3\frac{\mathrm{d}u_2}{\mathrm{d}x}\right)v_1\right]_0^{\pi}+\int_0^{\pi}\frac{\mathrm{d}u_1}{\mathrm{d}x}\frac{\mathrm{d}v_1}{\mathrm{d}x}+3\frac{\mathrm{d}u_2}{\mathrm{d}x}\frac{\mathrm{d}v_1}{\mathrm{d}x}-u_1v_1+8u_2v_1\,\mathrm{d}x=$$
$$\int_0^{\pi}20v_1\sin 2x\,\mathrm{d}x.$$

Applying the Neumann boundary condition given by Eq. (6.3) at $x = 0$, and the condition on v_1 given by Eq. (6.7) allows us to write the equation above as

$$\int_0^{\pi}\frac{\mathrm{d}u_1}{\mathrm{d}x}\frac{\mathrm{d}v_1}{\mathrm{d}x}+3\frac{\mathrm{d}u_2}{\mathrm{d}x}\frac{\mathrm{d}v_1}{\mathrm{d}x}-u_1v_1+8u_2v_1\,\mathrm{d}x=\int_0^{\pi}20v_1\sin 2x\,\mathrm{d}x-7v_1(0).$$

In the same spirit, we: (i) multiply Eq. (6.2) by the test function $v_2(x)$; (ii) integrate the resulting product over $0 < x < \pi$; (iii) integrate by parts; (iv) apply the Neumann boundary condition given by Eq. (6.6) at $x = \pi$; and (v) apply the condition on $v_2(x)$ given by Eq. (6.8). We then obtain

$$\int_0^{\pi}\frac{\mathrm{d}u_2}{\mathrm{d}x}\frac{\mathrm{d}v_2}{\mathrm{d}x}+2u_1v_2-4u_2v_2\,\mathrm{d}x=\int_0^{\pi}2v_2\sin x\,\mathrm{d}x+2v_2(\pi).$$

All that is now required to complete the weak formulation of the problem is to demand that $u_1(x)$ satisfies the Dirichlet boundary condition given by Eq. (6.5) and that $u_2(x)$ satisfies the Dirichlet boundary condition given by Eq. (6.4). We may then write the weak formulation of Eqs. (6.1) and (6.2), subject to the boundary conditions given by Eqs. (6.3)–(6.6) as:

find $u_1(x), u_2(x) \in H_E^1(0,\pi)$ such that

$$\int_0^{\pi}\frac{\mathrm{d}u_1}{\mathrm{d}x}\frac{\mathrm{d}v_1}{\mathrm{d}x}+3\frac{\mathrm{d}u_2}{\mathrm{d}x}\frac{\mathrm{d}v_1}{\mathrm{d}x}-u_1v_1+8u_2v_1\,\mathrm{d}x=\int_0^{\pi}20v_1\sin 2x\,\mathrm{d}x-7v_1(0), \quad (6.9)$$
$$\int_0^{\pi}\frac{\mathrm{d}u_2}{\mathrm{d}x}\frac{\mathrm{d}v_2}{\mathrm{d}x}+2u_1v_2-4u_2v_2\,\mathrm{d}x=\int_0^{\pi}2v_2\sin x\,\mathrm{d}x+2v_2(\pi). \quad (6.10)$$

for all $v_1(x), v_2(x) \in H_0^1(0,\pi)$.

6.3 The Mesh and Basis Functions

We will calculate a linear approximation to both $u_1(x)$ and $u_2(x)$ on each element. As such, we will use the mesh of N elements described in Sect. 3.3.1, and the linear basis functions described in Sect. 3.3.2.

6.4 The Finite Element Solution

Before we define the finite element solution, we need to specify the set S^h, and the subsets S_E^h and S_0^h of this set. The set S^h is defined to be the set of all functions $V^{(1)}(x)$, $V^{(2)}(x)$ that may be written as linear sums of the basis functions:

$$V^{(1)}(x) = \sum_{j=1}^{N+1} V_j^{(1)} \phi_j(x),$$

$$V^{(2)}(x) = \sum_{j=1}^{N+1} V_j^{(2)} \phi_j(x).$$

The set S_E^h is, as usual, defined to be the subset of S^h that contains functions that satisfy the Dirichlet boundary conditions. On using Eqs. (6.4) and (6.5), we see that these are the functions $V^{(1)}(x)$ and $V^{(2)}(x)$ that satisfy

$$V^{(1)}(\pi) = 0, \qquad V^{(2)}(0) = 0.$$

The set S_0^h is the subset of S^h that contains functions that take the value zero where Dirichlet boundary conditions are applied. As our model problem is the special case where the Dirichlet boundary conditions, given by Eqs. (6.4) and (6.5), set the solution to zero, the set S_0^h is identical to the set S_E^h.

We now define the finite element solution by

Find $U^{(1)}(x), U^{(2)}(x) \in S_E^h$ such that

$$\int_0^{\pi} \frac{\mathrm{d}U^{(1)}}{\mathrm{d}x}\frac{\mathrm{d}V^{(1)}}{\mathrm{d}x} + 3\frac{\mathrm{d}U^{(2)}}{\mathrm{d}x}\frac{\mathrm{d}V^{(1)}}{\mathrm{d}x} - U^{(1)}V^{(1)} + 8U^{(2)}V^{(1)}\,\mathrm{d}x = \int_0^{\pi} 20V^{(1)}\sin 2x\,\mathrm{d}x - 7V^{(1)}(0), \tag{6.11}$$

$$\int_0^{\pi} \frac{\mathrm{d}U^{(2)}}{\mathrm{d}x}\frac{\mathrm{d}V^{(2)}}{\mathrm{d}x} + 2U^{(1)}V^{(2)} - 4U^{(2)}V^{(2)}\,\mathrm{d}x = \int_0^{\pi} 2V^{(2)}\sin x\,\mathrm{d}x + 2V^{(2)}(\pi). \tag{6.12}$$

for all $V^{(1)}(x), V^{(2)}(x) \in S_0^h$.

Note the small change in the convention for naming the finite element solutions—it might be thought that these finite element solutions would be named $U_1(x)$ and $U_2(X)$. We use superscripts, rather than subscripts, for these finite element solutions so that, for example, $U^{(1)}$ is not confused with U_1, the first entry of the vector $\mathbf{U}$.

6.5 The Algebraic Equations

We write the finite element solutions, $U^{(1)}(x)$ and $U^{(2)}(x)$, as members of the set S^h given by

$$U^{(1)}(x) = \sum_{j=1}^{N+1} U_j^{(1)} \phi_j(x), \tag{6.13}$$

$$U^{(2)}(x) = \sum_{j=1}^{N+1} U_j^{(2)} \phi_j(x), \tag{6.14}$$

where $U_j^{(1)}, U_j^{(2)}, j = 1, 2, \ldots, N+1$, are to be determined. As usual, the system of algebraic equations contains both: (i) equations that ensure that $U^{(1)}(x)$ and $U^{(2)}(x)$ satisfy the Dirichlet boundary conditions and are therefore members of the set S_E^h; and (ii) equations that arise through substituting suitable test functions into Eqs. (6.11) and (6.12).

The Dirichlet boundary conditions, given by Eqs. (6.4) and (6.5), require that

$$U^{(1)}(\pi) = 0, \qquad U^{(2)}(0) = 0.$$

We note that $x_{N+1} = \pi$ and $x_1 = 0$. From the definition of the finite element solutions given by Eqs. (6.13) and (6.14), we see that these boundary conditions are satisfied if

$$U_{N+1}^{(1)} = 0, \tag{6.15}$$

$$U_1^{(2)} = 0. \tag{6.16}$$

Eqs. (6.15) and (6.16) then specify our first two algebraic equations.

Test functions $V^{(1)}(x)$ for Eq. (6.11) must satisfy $V^{(1)}(x_{N+1}) = 0$ so that they lie in the space S_0^h. The N basis functions $\phi_i(x), i = 1, 2, \ldots, N$, satisfy this condition and are therefore suitable candidates. Substituting $V^{(1)}(x) = \phi_i(x), i = 1, 2, \ldots, N$, into Eq. (6.11), together with the finite element solutions given by Eqs. (6.13) and (6.14), yields, after some manipulation,

$$\sum_{j=1}^{N+1} \left(\int_0^{\pi} \frac{\mathrm{d}\phi_i}{\mathrm{d}x}\frac{\mathrm{d}\phi_j}{\mathrm{d}x} - \phi_i\phi_j \,\mathrm{d}x \right) U_j^{(1)} + \\ \sum_{j=1}^{N+1} \left(\int_0^{\pi} 3\frac{\mathrm{d}\phi_i}{\mathrm{d}x}\frac{\mathrm{d}\phi_j}{\mathrm{d}x} + 8\phi_i\phi_j \,\mathrm{d}x \right) U_j^{(2)} = \int_0^{\pi} 20\phi_i \sin 2x \,\mathrm{d}x - 7\phi_i(x_1).$$

This may be written

$$\sum_{j=1}^{N+1} A_{ij}^{(11)} U_j^{(1)} + \sum_{j=1}^{N+1} A_{ij}^{(12)} U_j^{(2)} = b_i^{(1)}, \qquad i = 1, 2, \ldots, N, \tag{6.17}$$

where for $i = 1, 2, \ldots, N$, $j = 1, 2, \ldots, N + 1$,

$$A_{ij}^{(11)} = \int_0^{\pi} \frac{\mathrm{d}\phi_i}{\mathrm{d}x}\frac{\mathrm{d}\phi_j}{\mathrm{d}x} - \phi_i\phi_j \,\mathrm{d}x, \tag{6.18}$$

$$A_{ij}^{(12)} = \int_0^{\pi} 3\frac{\mathrm{d}\phi_i}{\mathrm{d}x}\frac{\mathrm{d}\phi_j}{\mathrm{d}x} + 8\phi_i\phi_j \,\mathrm{d}x, \tag{6.19}$$

and

$$b_1^{(1)} = \int_0^{\pi} 20\phi_1 \sin 2x \,\mathrm{d}x - 7, \tag{6.20}$$

$$b_i^{(1)} = \int_0^{\pi} 20\phi_i \sin 2x \,\mathrm{d}x, \qquad i = 2, 3, \ldots, N. \tag{6.21}$$

We now combine Eqs. (6.15) and (6.17) into the single matrix equation

$$A^{(11)}\mathbf{U}^{(1)} + A^{(12)}\mathbf{U}^{(2)} = \mathbf{b}^{(1)}, \tag{6.22}$$

where the entries of $A^{(11)}$, $A^{(12)}$, $\mathbf{b}^{(1)}$ are given by Eqs. (6.18)–(6.21), together with

$$A_{N+1,j}^{(11)} = \begin{cases} 1, & j = N+1, \\ 0, & j \neq N+1, \end{cases} \tag{6.23}$$

$$A_{N+1,j}^{(12)} = 0, \qquad j = 1, 2, \ldots, N+1, \tag{6.24}$$

$$b_{N+1}^{(1)} = 0. \tag{6.25}$$

We now consider the integral given by Eq. (6.12). Suitable test functions will satisfy $V^{(2)}(x_1) = 0$, and this condition is satisfied by the N basis functions $\phi_i(x)$, $i = 2, 3, \ldots, N + 1$. Substituting these basis functions, and Eqs. (6.13) and (6.14), into Eq. (6.12) we obtain, in a similar fashion to above:

$$\sum_{j=1}^{N+1} A_{ij}^{(21)} U_j^{(1)} + \sum_{j=1}^{N+1} A_{ij}^{(22)} U_j^{(2)} = b_i^{(2)}, \qquad i = 2, 3, \ldots, N+1, \tag{6.26}$$

where for $i = 2, 3, \ldots, N + 1$, $j = 1, 2, \ldots, N + 1$,

$$A_{ij}^{(21)} = \int_0^{\pi} 2\phi_i\phi_j \,\mathrm{d}x, \tag{6.27}$$

$$A_{ij}^{(22)} = \int_0^{\pi} \frac{\mathrm{d}\phi_i}{\mathrm{d}x}\frac{\mathrm{d}\phi_j}{\mathrm{d}x} - 4\phi_i\phi_j \,\mathrm{d}x, \tag{6.28}$$

and

$$b_i^{(2)} = \int_0^{\pi} 2\phi_i \sin x \, \mathrm{d}x, \qquad i = 2, 3, \ldots, N, \tag{6.29}$$

$$b_{N+1}^{(2)} = \int_0^{\pi} 2\phi_{N+1} \sin x \, \mathrm{d}x + 2. \tag{6.30}$$

As before, Eqs. (6.16) and (6.26) may be written as the single equation

$$A^{(21)}\mathbf{U}^{(1)} + A^{(22)}\mathbf{U}^{(2)} = \mathbf{b}^{(2)}, \tag{6.31}$$

where the entries of the $A^{(21)}$, $A^{(22)}$, $\mathbf{b}^{(2)}$ are given by Eqs. (6.27)–(6.30), together with

$$A_{1,j}^{(21)} = 0, \qquad j = 1, 2, \ldots, N+1, \tag{6.32}$$

$$A_{1,j}^{(22)} = \begin{cases} 1, & j = 1, \\ 0, & j \neq 1, \end{cases} \tag{6.33}$$

$$b_1^{(2)} = 0. \tag{6.34}$$

Finally, we may combine Eqs. (6.22) and (6.31) into the single matrix equation

$$A\mathbf{U} = \mathbf{b}, \tag{6.35}$$

where

$$A = \begin{pmatrix} A^{(11)} & A^{(12)} \\ A^{(21)} & A^{(22)} \end{pmatrix}, \qquad \mathbf{U} = \begin{pmatrix} \mathbf{U}^{(1)} \\ \mathbf{U}^{(2)} \end{pmatrix}, \qquad \mathbf{b} = \begin{pmatrix} \mathbf{b}^{(1)} \\ \mathbf{b}^{(2)} \end{pmatrix}. \tag{6.36}$$

The values $U_1^{(1)}, U_2^{(1)}, \ldots, U_{N+1}^{(1)}$ and $U_1^{(2)}, U_2^{(2)}, \ldots, U_{N+1}^{(2)}$ that specify the finite element solution for $U^{(1)}(x)$ and $U^{(2)}(x)$ given by Eqs. (6.13) and (6.14) may then be obtained by solving the linear system given by Eq. (6.35).

6.6 Assembling the Algebraic Equations

We will assemble the global matrix A and global vector $\mathbf{b}$ that appear in the linear system Eq. (6.35) by assembling the submatrices and subvectors that A and $\mathbf{b}$ are comprised of, as shown in Eq. (6.36).

We start by assembling the matrix $A^{(11)}$ that is defined by Eqs. (6.18) and (6.23). We see that these entries take the same form as those calculated in earlier chapters: most entries (those defined by Eq. (6.18)) take the form of integrals of products of basis functions and their derivatives over the whole domain; other entries (those defined by Eq. (6.23)) take the form of very simple, explicit numerical values. We may therefore assemble $A^{(11)}$ using the techniques developed in earlier chapters. To do this, we first evaluate the entries that are defined by integrals, *i.e.* those given by Eq. (6.18). These are assembled by decomposing these integrals into the sum of integrals over individual elements. We loop over all elements, calculating only the nonzero local contributions to the entries of $A^{(11)}$ from each element, before adding these local contributions into $A^{(11)}$. Having done this, we then complete the assembly of this matrix by setting the entries defined by Eq. (6.23).

The other matrices $A^{(12)}$, $A^{(21)}$, $A^{(22)}$ may be assembled using the same approach to that described above for assembling $A^{(11)}$. Finally, the vectors $\mathbf{b}^{(1)}$ and $\mathbf{b}^{(2)}$ may also be assembled in a similar manner: we do, however, have an additional term in Eqs. (6.20) and (6.30) to add into these vectors that arise from the Neumann boundary conditions.

6.7 Computational Implementation

We now develop an example MATLAB implementation, given in Listing 6.1, of the material presented in this chapter. The function written requires that the mesh is specified by a user-supplied vector $\mathbf{x}$. We require that this mesh is consistent with the mesh used in this chapter. As such, if there are N elements in the mesh, then the vector $\mathbf{x}$ must be of length $N + 1$ with: (i) the first entry equal to zero, (ii) the last entry equal to π and (iii) the entries of the vector ordered in ascending order. This ensures that the vector $\mathbf{x}$ is a suitable partitioning of the domain $0 < x < \pi$. The function returns the vectors corresponding to $\mathbf{U}^{(1)}$ and $\mathbf{U}^{(2)}$. This function can be called by typing, for example,

```
x = linspace(0, pi, 101);
[U1, U2] = Chap6_CalculateFemForSystems(x);
```

into a MATLAB session.

As already noted, the vector $\mathbf{x}$ will be of length $N + 1$, where N is the number of elements. The length of $\mathbf{x}$ is used to deduce the number of elements in line 4 of the listing. The matrices $A^{(11)}$, $A^{(12)}$, $A^{(21)}$, $A^{(22)}$, and the vectors $\mathbf{b}^{(1)}$, $\mathbf{b}^{(2)}$ are then initialised to zero, and to be of the correct size, in lines 7–12.

In lines 16–28, we loop over all elements, calculating the local contributions to $A^{(11)}$, $A^{(12)}$, $A^{(21)}$, $A^{(22)}$, $\mathbf{b}^{(1)}$, $\mathbf{b}^{(2)}$, and then incrementing these global matrices and vectors appropriately.

The function used to calculate the local entries is given in lines 66–101. In this function, we first calculate the length of each element in line 71. We calculate the local contributions to $A^{(11)}$, $A^{(12)}$, $A^{(21)}$, $A^{(22)}$ analytically in lines 74–79. We then define a quadrature rule in lines 82–84 that we use to calculate the entries of $\mathbf{b}^{(1)}$, $\mathbf{b}^{(2)}$ in lines 92–101.

Returning to the main body of code, we handle the Neumann boundary conditions in lines 31 and 34 and the Dirichlet boundary conditions in lines 37–46. We then form the global matrix A in line 49, and the global vector $\mathbf{b}$ in line 50, before solving the equation in line 53. The vectors $\mathbf{U}_1$ and $\mathbf{U}_2$ are extracted from the solution of the linear system in lines 56 and 57. Finally, in lines 60–62, we plot the finite element solution given by both $U^{(1)}(x)$ and $U^{(2)}(x)$.

Listing 6.1 `Chap6_CalculateFemForSystems.m`, a MATLAB function for calculating the finite element solution of the system of differential equations given by Eqs. (6.1)–(6.6).

```
1  function [U1, U2] = Chap6_CalculateFemForSystems(x)
2
3  % Deduce the number of elements
4  N = length(x) - 1;
5
6  % Initialise A_11, A_12, A_21, A_22, b_1, b_2 to zero
7  A_11 = zeros(N+1, N+1);
8  A_12 = zeros(N+1, N+1);
9  A_21 = zeros(N+1, N+1);
10 A_22 = zeros(N+1, N+1);
11 b_1 = zeros(N+1, 1);
12 b_2 = zeros(N+1, 1);
13
14 % Loop over elements, calculating nonzero local
15 % contributions and inserting them into linear system
16 for k=1:N
17     [A_local_11, A_local_12, A_local_21, ...
18         A_local_22, b_local_1, b_local_2] = ...
19         CalculateContributionOnElement(x, k);
20
21     A_11(k:k+1, k:k+1) = A_11(k:k+1, k:k+1) + A_local_11;
22     A_12(k:k+1, k:k+1) = A_12(k:k+1, k:k+1) + A_local_12;
23     A_21(k:k+1, k:k+1) = A_21(k:k+1, k:k+1) + A_local_21;
24     A_22(k:k+1, k:k+1) = A_22(k:k+1, k:k+1) + A_local_22;
25
26     b_1(k:k+1) = b_1(k:k+1) + b_local_1;
27     b_2(k:k+1) = b_2(k:k+1) + b_local_2;
28 end
29
30 % Handle Neumann boundary condition at x=0
31 b_1(1) = b_1(1) - 7;
32
33 % Handle Neumann boundary condition at x=pi
34 b_2(N+1) = b_2(N+1) + 2;
35
36 % Handle Dirichlet boundary condition at x=pi
37 A_11(N+1,:) = 0;
38 A_11(N+1,N+1) = 1;
39 A_12(N+1,:) = 0;
40 b_1(N+1) = 0;
41
42 % Handle Dirichlet boundary condition at x=0
43 A_21(1,:) = 0;
44 A_22(1,:) = 0;
45 A_22(1,1) = 1;
46 b_2(1) = 0;
47
48 % Form global matrix A and vector b
49 A = [A_11, A_12; A_21, A_22];
50 b = [b_1; b_2];
51
52 % Solve linear system
53 U = A\b;
54
55 % Decompose the solution U into U_1 and U_2
```

```
U1 = U(1:N+1);
U2 = U(N+2:2*N+2);

% Plot the solution
plot(x,U1,'r-',x,U2,'k-')
xlabel('x')
legend('U_1','U_2')



function [A_local_11, A_local_12, A_local_21, ...
    A_local_22, b_local_1, b_local_2] = ...
    CalculateContributionOnElement(x, k)

% Calculate length of element
h = x(k+1)-x(k);

% Calculate analytical contributions to A
A_local_11 = [1/h, -1/h; -1/h, 1/h] - [h/3, h/6; h/6, h/3];
A_local_12 = 3*[1/h, -1/h; -1/h, 1/h] + ...
    8*[h/3, h/6; h/6, h/3];
A_local_21 = 2*[h/3, h/6; h/6, h/3];
A_local_22 = [1/h, -1/h; -1/h, 1/h] - ...
    4*[h/3, h/6; h/6, h/3];

% Define quadrature rule
M = 3;
QuadWeights = [5/18, 4/9, 5/18];
QuadPoints = [0.5*(1-sqrt(0.6)), 0.5, 0.5*(1+sqrt(0.6))];

% Initialise local contributions to b to zero
b_local_1 = zeros(2,1);
b_local_2 = zeros(2,1);

% Loop over quadrature points to evaluate local
% contributions to b
for m = 1:M
    X = QuadPoints(m);
    phi_local = [1-X, X];
    for i=1:2
        b_local_1(i) = b_local_1(i) + h*QuadWeights(m)* ...
            20*sin(2*(x(k)+h*X))*phi_local(i);
        b_local_2(i) = b_local_2(i) + h*QuadWeights(m)* ...
            2*sin(x(k)+h*X)*phi_local(i);
    end
end
```

6.8 More General Linear Problems

We now give a brief summary of how to calculate the finite element solution of a system of N_{eq} ordinary differential equations. Suppose the unknown functions are $u_1(x), u_2(x), \ldots, u_{N_{eq}}(x)$. Using vector notation, we may write

$$\mathbf{u}(x) = \begin{pmatrix} u_1(x) \\ u_2(x) \\ \vdots \\ u_{N_{\text{eq}}}(x) \end{pmatrix}, \qquad \frac{\mathrm{d}\mathbf{u}}{\mathrm{d}x} = \begin{pmatrix} \frac{\mathrm{d}u_1}{\mathrm{d}x} \\ \frac{\mathrm{d}u_2}{\mathrm{d}x} \\ \vdots \\ \frac{\mathrm{d}u_{N_{\text{eq}}}}{\mathrm{d}x} \end{pmatrix}, \qquad \frac{\mathrm{d}^2\mathbf{u}}{\mathrm{d}x^2} = \begin{pmatrix} \frac{\mathrm{d}^2u_1}{\mathrm{d}x^2} \\ \frac{\mathrm{d}^2u_2}{\mathrm{d}x^2} \\ \vdots \\ \frac{\mathrm{d}^2u_{N_{\text{eq}}}}{\mathrm{d}x^2} \end{pmatrix}.$$

A general linear, second-order system of differential equations may then be written, for $a < x < b$, as

$$-\frac{\mathrm{d}}{\mathrm{d}x}\left(P(x)\frac{\mathrm{d}\mathbf{u}}{\mathrm{d}x}\right) + Q(x)\frac{\mathrm{d}\mathbf{u}}{\mathrm{d}x} + R(x)\mathbf{u} = \mathbf{f}(x), \tag{6.37}$$

where $P(x)$, $Q(x)$, $R(x)$ are matrices of size $N_{\text{eq}} \times N_{\text{eq}}$, and $\mathbf{f}$ is a vector of length N_{eq}. Alternatively, we may write Eq. (6.37) in component form as, for $m = 1, 2, \ldots, N_{\text{eq}}$, $a < x < b$,

$$-\sum_{n=1}^{N_{\text{eq}}} \frac{\mathrm{d}}{\mathrm{d}x}\left(P_{mn}(x)\frac{\mathrm{d}u_n}{\mathrm{d}x}\right) + \sum_{n=1}^{N_{\text{eq}}} Q_{mn}(x)\frac{\mathrm{d}u_n}{\mathrm{d}x} + \sum_{n=1}^{N_{\text{eq}}} R_{mn}(x)u_n = f_m(x). \tag{6.38}$$

We assume that each equation has one boundary condition prescribed at $x = a$, and one boundary condition prescribed at $x = b$.

6.8.1 The Weak Formulation

The boundary conditions for the system of differential equations given by Eq. (6.38) allow us to define the Sobolev spaces $H_E^1(a, b)$ and $H_0^1(a, b)$ for this system of equations. Multiplying each component $m = 1, 2, \ldots, N_{\text{eq}}$ of the differential equation by a suitable test function $v_m(x)$, integrating the resulting product over the domain $a < x < b$ and integrating the first term by parts, we may then write the weak formulation of the differential equation as:

find $u_1(x), u_2(x), \ldots, u_{N_{\text{eq}}}(x) \in H_E^1(a, b)$ such that, for $m = 1, 2, \ldots, N_{\text{eq}}$:

$$\int_a^b \sum_{n=1}^{N_{\text{eq}}} \left(P_{mn}\frac{\mathrm{d}u_n}{\mathrm{d}x}\frac{\mathrm{d}v_m}{\mathrm{d}x} + Q_{mn}\frac{\mathrm{d}u_n}{\mathrm{d}x}v_m + R_{mn}u_n v_m\right)\mathrm{d}x =$$

$$\int_a^b f_m v_m \,\mathrm{d}x + h_m(v_m(a), v_m(b)),$$

for all $v_m(x) \in H_0^1(a, b)$,

where $h_m(v_m(a), v_m(b))$ are the contributions from any Neumann boundary conditions satisfied by component m of the differential equation.

6.8.2 The Finite Element Solution

After specifying a suitable mesh, and defining basis functions on this mesh, we define S^h to be the set of vector-valued functions, with N_{eq} components, where each component may be written as a linear sum of these basis functions, that is,

$$V^{(m)}(x) = \sum_{j=1}^{N_{\text{node}}} V_j^{(m)} \phi_j(x), \qquad m = 1, 2, \ldots, N_{\text{eq}}, \tag{6.39}$$

where N_{node} is the number of nodes in the mesh. The sets S_E^h and S_0^h are then subsets of S^h, with definitions that depend on the boundary conditions used. Having defined the sets S_E^h and S_0^h, we then define the finite element solution by:

find $U^{(1)}(x), U^{(2)}(x), \ldots, U^{(N_{\text{eq}})}(x) \in S_E^h$ such that, for $m = 1, 2, \ldots, N_{\text{eq}}$:

$$\int_a^b \sum_{n=1}^{N_{\text{eq}}} \left(P_{mn} \frac{\mathrm{d}U^{(n)}}{\mathrm{d}x} \frac{\mathrm{d}V^{(m)}}{\mathrm{d}x} + Q_{mn} \frac{\mathrm{d}U^{(n)}}{\mathrm{d}x} V^{(m)} + R_{mn} U^{(n)} V^{(m)} \, \mathrm{d}x \right) = \int_a^b f_m V^{(m)} \, \mathrm{d}x + h_m(V^{(m)}(a), V^{(m)}(b)), \tag{6.40}$$

for all $V^{(m)}(x) \in S_0^h$.

We now write the finite element solution as a member of S^h:

$$U^{(m)}(x) = \sum_{j=1}^{N_{\text{node}}} U_j^{(m)} \phi_j(x), \qquad m = 1, 2, \ldots, N_{\text{eq}}. \tag{6.41}$$

As usual, for each differential equation, a system of N_{node} algebraic equations may be derived by: (i) ensuring that $U^{(m)}(x) \in S_E^h$; and (ii) substituting $V^{(m)} = \phi_i(x)$ into Eq. (6.40) for suitable test functions $\phi_i(x) \in S_0^h$. Using a similar approach to that described in Sect. 6.5, this results in a linear system that may be written in the form $A\mathbf{U} = \mathbf{b}$ where

$$A = \begin{pmatrix} A^{(11)} & A^{(12)} & \ldots & A^{(1,N_{\text{eq}})} \\ A^{(21)} & A^{(22)} & \ldots & A^{(2,N_{\text{eq}})} \\ \vdots & \vdots & \ddots & \vdots \\ A^{(N_{\text{eq}},1)} & A^{(N_{\text{eq}},2)} & \ldots & A^{(N_{\text{eq}},,N_{\text{eq}})} \end{pmatrix},$$

$$\mathbf{U} = \begin{pmatrix} \mathbf{U}^{(1)} \\ \mathbf{U}^{(2)} \\ \vdots \\ \mathbf{U}^{(N_{\text{eq}})} \end{pmatrix},$$

$$\mathbf{b} = \begin{pmatrix} \mathbf{b}^{(1)} \\ \mathbf{b}^{(2)} \\ \vdots \\ \mathbf{b}^{(N_{\text{eq}})} \end{pmatrix},$$

We then solve the linear system $A\mathbf{U} = \mathbf{b}$ to evaluate the coefficients in the finite element solution given by Eq. (6.41).

6.9 Nonlinear Problems

We conclude this chapter with a brief summary of how the material in Sects. 6.1–6.7 may be adapted for nonlinear boundary value problems. Suppose we are solving the nonlinear system defined by, for $a < x < b$,

$$-\frac{\mathrm{d}}{\mathrm{d}x}\left(p_1\left(x, u_1, u_2, \frac{\mathrm{d}u_1}{\mathrm{d}x}, \frac{\mathrm{d}u_2}{\mathrm{d}x}\right)\right) + r_1\left(x, u_1, u_2, \frac{\mathrm{d}u_1}{\mathrm{d}x}, \frac{\mathrm{d}u_2}{\mathrm{d}x}\right) = 0,$$
$$-\frac{\mathrm{d}}{\mathrm{d}x}\left(p_2\left(x, u_1, u_2, \frac{\mathrm{d}u_1}{\mathrm{d}x}, \frac{\mathrm{d}u_2}{\mathrm{d}x}\right)\right) + r_2\left(x, u_1, u_2, \frac{\mathrm{d}u_1}{\mathrm{d}x}, \frac{\mathrm{d}u_2}{\mathrm{d}x}\right) = 0,$$

subject to suitable prescribed boundary conditions. Using the usual arguments, we find that the weak formulation is given by:

find $u_1(x), u_2(x) \in H_E^1(a, b)$ such that

$$\int_a^b p_1\left(x, u_1, u_2, \frac{\mathrm{d}u_1}{\mathrm{d}x}, \frac{\mathrm{d}u_2}{\mathrm{d}x}\right)\frac{\mathrm{d}v_1}{\mathrm{d}x} + r_1\left(x, u_1, u_2, \frac{\mathrm{d}u_1}{\mathrm{d}x}, \frac{\mathrm{d}u_2}{\mathrm{d}x}\right) v_1(x)\,\mathrm{d}x + h_1(v_1(a), v_1(b)) = 0,$$

$$\int_a^b p_2\left(x, u_1, u_2, \frac{\mathrm{d}u_1}{\mathrm{d}x}, \frac{\mathrm{d}u_2}{\mathrm{d}x}\right)\frac{\mathrm{d}v_2}{\mathrm{d}x} + r_2\left(x, u_1, u_2, \frac{\mathrm{d}u_1}{\mathrm{d}x}, \frac{\mathrm{d}u_2}{\mathrm{d}x}\right) v_2(x)\,\mathrm{d}x + h_2(v_2(a), v_2(b)) = 0,$$

for all $v_1(x), v_2(x) \in H_0^1(a, b)$,

where $H_E^1(a, b)$ and $H_0^1(a, b)$ are defined using the boundary conditions for each equation, and $h_1(v_1(a), v_1(b))$ and $h_2(v_2(a), v_2(b))$ represent the contributions, if any, from Neumann boundary conditions.

Defining the computational mesh allows us to specify suitable basis functions and, in turn, the sets S_E^h and S_0^h. We then write the finite element solutions as the members of S^h given by:

$$U^{(1)}(x) = \sum_{j=1}^{N_{\text{node}}} U_j^{(1)} \phi_j(x), \qquad U^{(2)}(x) = \sum_{j=1}^{N_{\text{node}}} U_j^{(2)} \phi_j(x),$$

where N_{node} is the number of nodes in the mesh. The finite element solution is then defined by:

find $U^{(1)}(x), U^{(2)}(x) \in S_E^h$ such that

$$\begin{aligned}\int_a^b p_1 \left(x, U^{(1)}, U^{(2)}, \frac{\mathrm{d}U^{(1)}}{\mathrm{d}x}, \frac{\mathrm{d}U^{(2)}}{\mathrm{d}x}\right) \frac{\mathrm{d}V^{(1)}}{\mathrm{d}x} + \\ r_1 \left(x, U^{(1)}, U^{(2)}, \frac{\mathrm{d}U^{(1)}}{\mathrm{d}x}, \frac{\mathrm{d}U^{(2)}}{\mathrm{d}x}\right) V^{(1)}(x)\, \mathrm{d}x + \\ h_1(V^{(1)}(a), V^{(1)}(b)) = 0,\end{aligned} \tag{6.42}$$

$$\begin{aligned}\int_a^b p_2 \left(x, U^{(1)}, U^{(2)}, \frac{\mathrm{d}U^{(1)}}{\mathrm{d}x}, \frac{\mathrm{d}U^{(2)}}{\mathrm{d}x}\right) \frac{\mathrm{d}V^{(2)}}{\mathrm{d}x} + \\ r_2 \left(x, U^{(1)}, U^{(2)}, \frac{\mathrm{d}U^{(1)}}{\mathrm{d}x}, \frac{\mathrm{d}U^{(2)}}{\mathrm{d}x}\right) V^{(2)}(x)\, \mathrm{d}x + \\ h_2(V^{(2)}(a), V^{(2)}(b)) = 0,\end{aligned} \tag{6.43}$$

for all $V^{(1)}(x), V^{(2)}(x) \in S_0^h$.

Substituting $\phi_i(x)$ into Eq. (6.42) for all permissible $\phi_i(x)$, when combined with enforcing $U^{(1)}(x) \in S_E^h$, will result in N_{node} algebraic equations, which we write as

$$\mathbf{R}^{(1)} \left(U_1^{(1)}, U_2^{(1)}, \ldots, U_{N_{\text{node}}}^{(1)}, U_1^{(2)}, U_2^{(2)}, \ldots, U_{N_{\text{node}}}^{(2)}\right) = \mathbf{0}. \tag{6.44}$$

Similarly, substituting $\phi_i(x)$ into Eq. (6.43) for all permissible $\phi_i(x) \in S_0^h$, and enforcing $U^{(2)}(x) \in S_E^h$, will result in a further N_{node} algebraic equations,

$$\mathbf{R}^{(2)} \left(U_1^{(1)}, U_2^{(1)}, \ldots, U_{N_{\text{node}}}^{(1)}, U_1^{(2)}, U_2^{(2)}, \ldots, U_{N_{\text{node}}}^{(2)}\right) = \mathbf{0}. \tag{6.45}$$

The residuals $\mathbf{R}^{(1)}$ and $\mathbf{R}^{(2)}$ may be calculated in the same way as the residuals for the scalar value nonlinear differential equations discussed in Chap. 5.

The two systems of nonlinear algebraic equations defined by Eqs. (6.44) and (6.45) may then be combined into a single system of nonlinear algebraic equations given by:

$$\mathbf{R} = \mathbf{0}, \qquad \text{where} \qquad \mathbf{R} = \begin{pmatrix} \mathbf{R}^{(1)} \\ \mathbf{R}^{(1)} \end{pmatrix}.$$

Provided the Jacobian matrix is not required, this residual vector may be used to determine $U_1^{(1)}, U_2^{(1)}, \ldots, U_{N_{\text{node}}}^{(1)}, U_1^{(2)}, U_2^{(2)}, \ldots, U_{N_{\text{node}}}^{(2)}$. If the Jacobian is required, it may be written in the form

$$J = \begin{pmatrix} J^{(11)} & J^{(12)} \\ J^{(21)} & J^{(22)} \end{pmatrix},$$

where the entries of the submatrices of J are given by, for $i = 1, 2, \ldots, N_{\text{node}}$, $j = 1, 2, \ldots, N_{\text{node}}$,

$$J_{i,j}^{(11)} = \frac{\partial R_i^{(1)}}{\partial U_j^{(1)}}, \qquad J_{i,j}^{(12)} = \frac{\partial R_i^{(1)}}{\partial U_j^{(2)}},$$

$$J_{i,j}^{(21)} = \frac{\partial R_i^{(2)}}{\partial U_j^{(1)}}, \qquad J_{i,j}^{(22)} = \frac{\partial R_i^{(2)}}{\partial U_j^{(2)}}.$$

The entries of these submatrices may be calculated in the same way as the Jacobian matrix for scalar value nonlinear differential equations, described in Chap. 5.

6.10 Exercises

1 A boundary value problem for a system of two differential equations is defined by, for $0 < x < 1$,

$$-2\frac{\mathrm{d}^2u_1}{\mathrm{d}x^2} - \frac{\mathrm{d}^2u_2}{\mathrm{d}x^2} + 2u_1 + 5u_2 = 2\mathrm{e}^{2x},$$

$$\frac{\mathrm{d}^2u_1}{\mathrm{d}x^2} - 3\frac{\mathrm{d}^2u_2}{\mathrm{d}x^2} + 12u_2 = \mathrm{e}^{x},$$

subject to Dirichlet boundary conditions at $x = 0$ given by

$$u_1(0) = 1,$$
$$u_2(0) = 2,$$

and Neumann boundary conditions at $x = 1$ given by

$$2\frac{\mathrm{d}u_1}{\mathrm{d}x} + \frac{\mathrm{d}u_2}{\mathrm{d}x} = 2\mathrm{e} + 4\mathrm{e}^2,$$

$$\frac{\mathrm{d}u_1}{\mathrm{d}x} - 3\frac{\mathrm{d}u_2}{\mathrm{d}x} = \mathrm{e} - 12\mathrm{e}^2.$$

The first Neumann boundary condition is a natural boundary condition for the first differential equation, and the second Neumann boundary condition is a natural boundary condition for the second differential equation.

This boundary value problem has analytic solution $u_1(x) = \mathrm{e}^x$, $u_2(x) = 2\mathrm{e}^{2x}$.

(a) In this chapter, we wrote down a model boundary value problem in Sect. 6.1 that was similar to the system above. We then explained how to solve this system, providing a computational implementation in Sect. 6.7. Using a similar pattern to that presented for this model problem, define: (i) the weak formulation, (ii) the finite element solution that is a linear approximation to the solution on each element and (iii) the algebraic system of equations satisfied by the finite element solution. Then, develop a computational implementation for setting up and solving the system of algebraic equations and confirm that your solution agrees with the analytic solution.

(b) Develop the material presented in this chapter to allow computation of a finite element solution to the boundary value problem above, using a quadratic approximation to the solution on each element.

Chapter 7
Linear Elliptic Partial Differential Equations

In earlier chapters, we described how to apply the finite element method to ordinary differential equations. For the remainder of this book, we will focus on extending this technique for application to partial differential equations. As with ordinary differential equations, we begin with a simple example to illustrate the key features. We then discuss more complex differential equations in later chapters.

The material in this chapter requires some knowledge of vector calculus and integration over two-dimensional regions. We assume that the reader has some familiarity with this topic, including: (i) the definition of the gradient operator ∇; (ii) basic identities that use the gradient operator; (iii) use of the divergence theorem; and (iv) surface and line integrals. A very brief summary of some of the vector calculus required is given in Appendix B.

7.1 A First Model Problem

Our initial, simple model problem is the partial differential equation

$$-\left(\frac{\partial^2 u}{\partial x^2} + \frac{\partial^2 u}{\partial y^2}\right) = 1, \qquad 0 < x < 1, \quad 0 < y < 1, \tag{7.1}$$

subject to Dirichlet boundary conditions on the whole boundary, given by

$$u(x, 0) = 0, \qquad 0 < x < 1, \tag{7.2}$$

$$u(1, y) = 0, \qquad 0 < y < 1, \tag{7.3}$$

$$u(x, 1) = 0, \qquad 0 < x < 1, \tag{7.4}$$

$$u(0, y) = 0, \qquad 0 < y < 1. \tag{7.5}$$

Our strategy when applying the finite element method to this model problem will be the same as that used when applying the finite element method to ordinary

J. Whiteley, *Finite Element Methods*, Mathematical Engineering,
DOI 10.1007/978-3-319-49971-0_7

differential equations. A summary of this strategy was written down in Sect. 2.9. We encourage the reader to revisit this summary before reading the remainder of this chapter.

7.2 The Weak Formulation

As we have seen throughout this book, the first step when using the finite element method is to derive the weak formulation of the differential equation. As with ordinary differential equations, this is done by first multiplying the given differential equation by a function in an appropriate Sobolev subspace. Before we may do this for the model problem given in Sect. 7.1, we must define Sobolev spaces for functions of two variables that generalise the definitions given for functions of one variable in Sect. 3.2.1.

7.2.1 Sobolev Spaces in Two Dimensions

Let Ω be a subset of the (x, y)-plane. In Sect. 3.2.1, we defined the concept of square integrable functions for functions of one variable. This definition may easily be extended to functions of two variables. A function $v(x, y)$ that is defined on the region Ω is said to be square integrable on this region if the integral

$$\int_\Omega (v(x, y))^2 \, \mathrm{d}A$$

exists. The set of square integrable functions on this region, denoted by $L^2(\Omega)$, is then defined by

$$L^2(\Omega) = \left\{ \text{functions } v(x, y) \text{ such that the integral } \int_\Omega (v(x, y))^2 \, \mathrm{d}A \text{ exists} \right\}.$$

The Sobolev space of order 1 for functions defined on the region Ω, denoted by $H^1(\Omega)$, is now defined to be those functions $v(x, y)$ that (i) are defined and continuous on Ω; (ii) lie in the set $L^2(\Omega)$; and (iii) have first partial derivatives with respect to both x and y that lie in $L^2(\Omega)$. Hence,

$$H^1(\Omega) = \left\{ \text{functions } v(x, y) \text{ such that } v(x, y) \text{ is continuous, and the integrals} \right.$$
$$\left. \int_\Omega (v(x, y))^2 \, \mathrm{d}A, \ \int_\Omega \left(\frac{\partial v}{\partial x}\right)^2 \mathrm{d}A, \text{ and } \int_\Omega \left(\frac{\partial v}{\partial y}\right)^2 \mathrm{d}A \text{ exist} \right\}. \tag{7.6}$$

We may also define the subsets $H_E^1(\Omega)$ and $H_0^1(\Omega)$ of $H^1(\Omega)$ that mirror those defined for functions of one variable in Sect. 3.2.1. Suppose a differential equation, together with suitable boundary conditions, has been specified. Using these boundary conditions, we define these two subsets by:

$$H_E^1(\Omega) = \Big\{\text{functions } v(x, y) \text{ in } H^1(\Omega) \text{ such that } v(x, y) \text{ satisfies all Dirichlet boundary conditions}\Big\}, \tag{7.7}$$

$$H_0^1(\Omega) = \Big\{\text{functions } v(x, y) \text{ in } H^1(\Omega) \text{ such that } v(x, y) = 0 \text{ at all points } (x, y) \text{ where Dirichlet boundary conditions are specified}\Big\}. \tag{7.8}$$

7.2.2 *Deriving the Weak Formulation*

The example differential equation, Eq. (7.1), is defined on the region $0 < x < 1$, $0 < y < 1$, which we will denote by Ω. This differential equation may be written in terms of the gradient operator ∇ as

$$-\nabla \cdot (\nabla u) = 1, \qquad (x, y) \in \Omega.$$

Multiplying this equation by any function $v(x, y) \in H_0^1(\Omega)$ gives

$$-v\nabla \cdot (\nabla u) = v, \qquad (x, y) \in \Omega.$$

After a little vector calculus—the use of Eq. (B.2) from Appendix B—this may be written

$$-\nabla \cdot (v\nabla u) + (\nabla v) \cdot (\nabla u) = v, \qquad (x, y) \in \Omega.$$

We now integrate this equation over the region Ω and apply the divergence theorem to the first term on the left-hand side. Denoting the boundary of Ω by $\partial\Omega$ yields

$$-\int_{\partial\Omega} v\,(\nabla u) \cdot \mathbf{n}\, \mathrm{d}s + \int_{\Omega} (\nabla v) \cdot (\nabla u)\, \mathrm{d}A = \int_{\Omega} v\, \mathrm{d}A, \tag{7.9}$$

where $\mathbf{n}$ denotes the normal vector to the boundary $\partial\Omega$ that points out of the region Ω and is of unit length. Note that, at this stage, some authors use the notation

$$\frac{\partial u}{\partial n} = (\nabla u) \cdot \mathbf{n}.$$

Remembering that $v(x, y) \in H_0^1(\Omega)$, the function $v(x, y)$ will take the value zero at all points on the boundary of Ω where Dirichlet boundary conditions are applied. The boundary conditions for our model problem, Eqs. (7.2)–(7.5), prescribe Dirichlet boundary conditions on the whole boundary $\partial\Omega$. Consequently, $v(x, y) = 0$ for every point on $\partial\Omega$. The first integral on the left-hand side of Eq. (7.9) will therefore be zero, and we may write

$$\int_\Omega (\nabla v) \cdot (\nabla u) \, \mathrm{d}A = \int_\Omega v \, \mathrm{d}A. \tag{7.10}$$

We are now ready to define the weak solution of our example differential equation. As with ordinary differential equations, the weak solution will satisfy both the Dirichlet boundary conditions and the integral condition given by Eq. (7.10), for all functions $v(x, y) \in H_0^1(\Omega)$. More precisely, the weak solution of Eqs. (7.1)–(7.5) is given by

find $u(x, y) \in H_E^1(\Omega)$ such that

$$\int_\Omega (\nabla v) \cdot (\nabla u) \, \mathrm{d}A = \int_\Omega v \, \mathrm{d}A, \tag{7.11}$$

for all $v(x, y) \in H_0^1(\Omega)$.

7.3 The Mesh and Basis Functions

Having derived the weak formulation of the differential equation, the next task is to partition the region on which the differential equation is specified into elements and to define the basis functions on these elements. In this chapter, we will partition Ω into a mesh of triangular elements. An alternative approach would be to partition Ω into quadrilateral elements; we will discuss this approach in Chap. 9. Both strategies for partitioning the domain have their merits, which are beyond the scope of this book. For most problems, however, either approach is suitable. We will consider only linear basis functions in this chapter; higher order basis functions will be introduced in Chap. 10.

7.3.1 A Mesh of Triangular Elements

Partitioning a general two-dimensional domain into a mesh of elements—known as *mesh generation*—is a nontrivial task. However, our model problem is defined on the square $0 < x < 1, 0 < y < 1$, and we will see that this region may be partitioned into triangular elements very easily. For more complex regions, established mesh generation packages exist. These will be discussed in Sect. 7.10.

The domain for the model problem, i.e. the square $0 < x < 1, 0 < y < 1$, may be partitioned into triangular elements by first partitioning the region into equally sized rectangles so that there are N_x rectangles in the x-direction and N_y rectangles in the y-direction. This is illustrated for $N_x = 2$ and $N_y = 3$ in Fig. 7.1a. We then partition each rectangle into two triangles by drawing a line across one diagonal of the rectangle, as illustrated in Fig. 7.1b. The elements of the mesh are then the triangles generated, and the nodes are the vertices of the elements.

We now number both the nodes and the elements of the mesh shown in Fig. 7.1b. When numbering nodes, we start with the node at the bottom left-hand corner of the mesh and number the nodes consecutively moving in the positive x-direction. We then move up to the next row in the y-direction and repeat this process, starting from the left-hand end of this row. We proceed through the mesh in this fashion until all nodes have been numbered. This numbering convention is illustrated in Fig. 7.2a when $N_x = 2$ and $N_y = 3$. A similar process is used to number the elements—this is illustrated in Fig. 7.2b when $N_x = 2$ and $N_y = 3$.

When implementing the finite element method later in this chapter, we will find it useful to have access to an array that contains the nodes of each element listed in an anticlockwise direction. This array is known as the *connectivity array*, and the

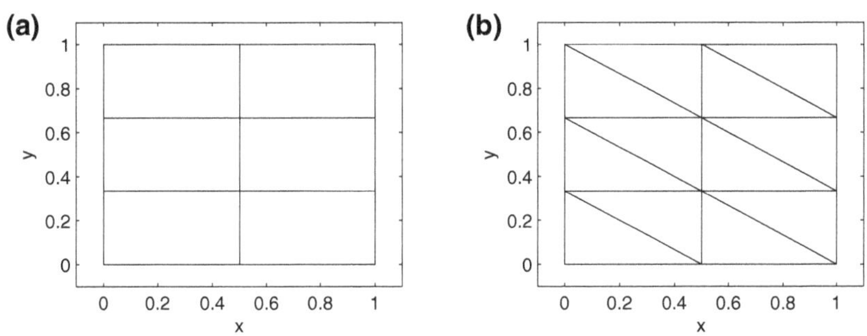

Fig. 7.1 **a** Partitioning of Ω into rectangles. **b** Partitioning of Ω into triangles

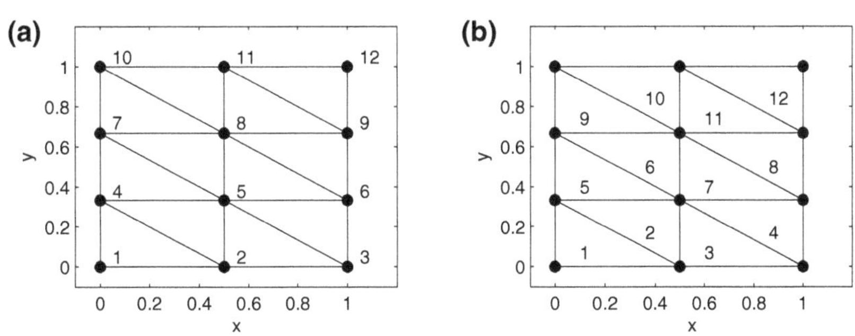

Fig. 7.2 **a** Labelling of the nodes in the mesh. **b** Labelling of the elements in the mesh

variable name traditionally used for this array in computer programs is `lnods`,[1] and we follow this convention here. Suppose our mesh contains N_{ele} elements. In matrix notation, $lnods_{k,i}$, where $k = 1, 2, \ldots, N_{\text{ele}}$ and $i = 1, 2, 3$, contains node i of element k. A suitable matrix $lnods$ for the mesh shown in Fig. 7.2 is given by

$$lnods = \begin{pmatrix} 1 & 2 & 4 \\ 5 & 4 & 2 \\ 2 & 3 & 5 \\ 6 & 5 & 3 \\ 4 & 5 & 7 \\ 8 & 7 & 5 \\ 5 & 6 & 8 \\ 9 & 8 & 6 \\ 7 & 8 & 10 \\ 11 & 10 & 8 \\ 8 & 9 & 11 \\ 12 & 11 & 9 \end{pmatrix}.$$

Note that the ordering of the nodes in each row of the connectivity array is not unique—we require only that the nodes of each element are listed in an anticlockwise order. The first row of the matrix $lnods$ above could, for example, be stored in the order $4, 1, 2$ or $2, 4, 1$. The reason for insisting that the elements are ordered in an anticlockwise direction will become apparent when we define basis functions in the next section. It is also important to note that the order in which the elements are listed in the matrix $lnods$ is not important. As such, we may interchange any rows of this matrix.

When assembling the system of algebraic equations that determine the finite element solution, we will require the value of x and y at each node. Suppose we have N_{node} nodes in the mesh. The values of x and y at node i, where $i = 1, 2, \ldots, N_{\text{node}}$, are denoted by x_i and y_i.

We now make some general comments on the partitioning of a domain into triangular elements. The partitioning of Ω illustrated in Fig. 7.2 has a key property that is required by any finite element mesh in more than one spatial dimension—any point in Ω that is a node of one element is also a node of any other element that it is contained in. For example, the partitioning of Ω shown in Fig. 7.3 is not an admissible partitioning, as a point exists—denoted by an open circle—that is a node of one element, but lies on the edge of another element without being a node of that element. This node is known as a *hanging node*.

---

[1] This variable name is a consequence of the design of early programming languages and their conventions for naming variables. Early versions of Fortran, the most common language of choice for scientific computing until fairly recently, used the convention that variables representing integers should have a variable name that starts with one of the letters `i`, `j`, `k`, `l`, `m`, `n`, or upper-case letters corresponding to these letters. Furthermore, variable names were restricted to a maximum of six characters.

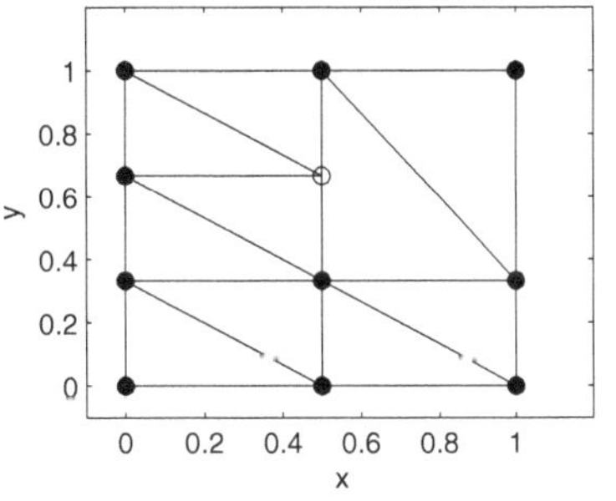

Fig. 7.3 An example of a hanging node

There is a further consideration to be aware of when using a mesh in two, or higher, dimensions. Long, thin elements should be avoided, as these elements can degrade the accuracy of the solution. There are many metrics for quantifying this property for an individual element, for example the minimum interior angle of an element, the maximum interior angle of an element, or the ratio of the length of the longest edge to the length of the shortest edge. It should be noted, however, that the loss of accuracy usually only becomes apparent in very extreme cases—for example when the minimum interior angle is less than around 10°.

7.3.2 *Linear Basis Functions*

In this chapter, we will calculate a finite element solution of the model problem that is a linear approximation to the true solution on each element. We now identify suitable basis functions for the mesh described in Sect. 7.3.1. We first define a canonical triangular element in (X, Y)-coordinates to be the triangle that occupies the region $X \geq 0, Y \geq 0, 1 - X - Y \geq 0$, as shown in Fig. 7.4. The global basis functions are then defined through (i) specification of local linear basis functions on this canonical triangular element; and (ii) a linear transformation that maps between an element of the mesh in (x, y)-coordinates and the canonical triangular element in (X, Y)-coordinates.

Suppose the nodes of element k in a mesh of triangular elements, ordered in an anticlockwise direction, are the nodes k_1, k_2 and k_3. For the mesh described in Sect. 7.3.1, these nodes will be the entries in row k of the matrix $lnods$, and the values of x and y at these nodes are given by $x_{k_1}, x_{k_2}, x_{k_3}$ and $y_{k_1}, y_{k_2}, y_{k_3}$. A mapping between this triangular element, in (x, y)-coordinates, and the canonical triangular element described above, in (X, Y)-coordinates, is given by

$$x = (1 - X - Y)x_{k_1} + Xx_{k_2} + Yx_{k_3}, \tag{7.12}$$

$$y = (1 - X - Y)y_{k_1} + Xy_{k_2} + Yy_{k_3}. \tag{7.13}$$

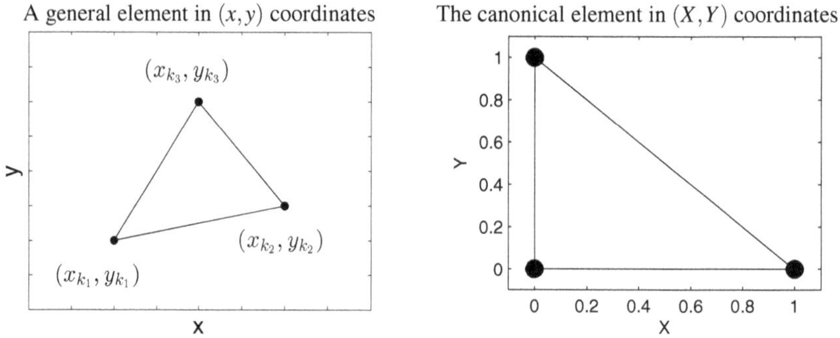

Fig. 7.4 Mapping between a general element in (x, y)-coordinates and the canonical element in (X, Y)-coordinates

This mapping is illustrated in Fig. 7.4. It is straightforward to verify that the transformation given by Eqs. (7.12) and (7.13) correctly maps the canonical element in (X, Y)-coordinates to element k in (x, y)-coordinates. The vertices of the canonical triangle, listed in an anticlockwise order, are the point $X = 0, Y = 0$, the point $X = 1, Y = 0$ and the point $X = 0, Y = 1$. These three points are clearly mapped, respectively, to nodes k_1, k_2 and k_3 of element k by Eqs. (7.12) and (7.13). Further, as the map is a linear function of X and Y, straight lines (such as the edges of a triangle) in the (X, Y)-coordinate system will be mapped to straight lines in the (x, y)-coordinate system.

It is now apparent why we insisted that the nodes of each element are listed in an anticlockwise order in the array *lnods*. If they were not, then the mapping between element k and the canonical element given by Eqs. (7.12) and (7.13) would include a reflection, which is not the intention of this transformation.

Local linear basis functions $\phi_{\text{local},1}(X, Y)$, $\phi_{\text{local},2}(X, Y)$, $\phi_{\text{local},3}(X, Y)$ are defined on the canonical element by

$$\phi_{\text{local},1}(X, Y) = 1 - X - Y, \tag{7.14}$$
$$\phi_{\text{local},2}(X, Y) = X, \tag{7.15}$$
$$\phi_{\text{local},3}(X, Y) = Y. \tag{7.16}$$

These local basis functions are illustrated in Fig. 7.5 and have the property that they take the value 1 at one node of the canonical triangle, and 0 at all other nodes:

$$\begin{aligned} &\phi_{\text{local},1}(0, 0) = 1, && \phi_{\text{local},1}(1, 0) = 0, && \phi_{\text{local},1}(0, 1) = 0, \\ &\phi_{\text{local},2}(0, 0) = 0, && \phi_{\text{local},2}(1, 0) = 1, && \phi_{\text{local},2}(0, 1) = 0, \\ &\phi_{\text{local},3}(0, 0) = 0, && \phi_{\text{local},3}(1, 0) = 0, && \phi_{\text{local},3}(0, 1) = 1. \end{aligned}$$

If there are N_{node} nodes in the mesh, there will be a total of N_{node} global basis functions ϕ_j, $j = 1, 2, \ldots, N_{\text{node}}$. We now define these global basis functions in terms

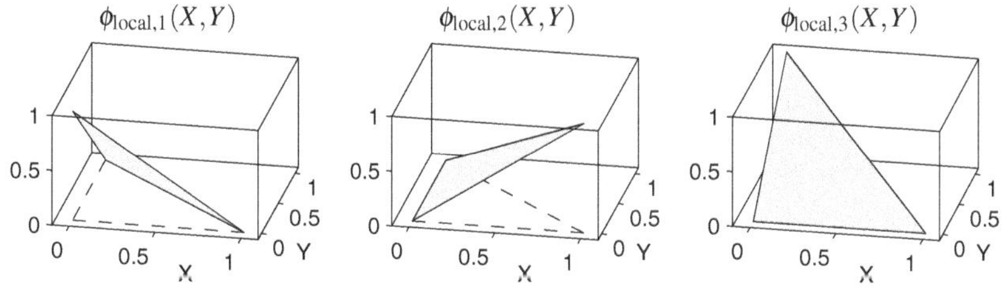

Fig. 7.5 Local basis functions $\phi_{\text{local},1}(X, Y)$, $\phi_{\text{local},2}(X, Y)$ and $\phi_{\text{local},3}(X, Y)$ defined by Eqs. (7.14)–(7.16)

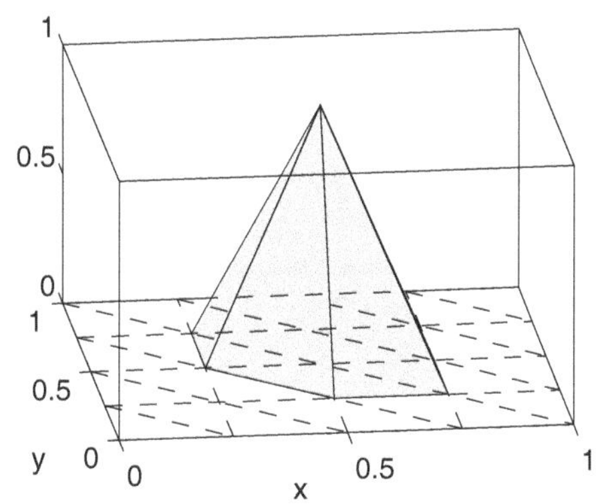

Fig. 7.6 An example basis function

of the local basis functions on each element. On each element $k = 1, 2, \ldots, N_{\text{ele}}$, we define

$$\phi_{k_1}(X, Y) = \phi_{\text{local},1}(X, Y), \tag{7.17}$$
$$\phi_{k_2}(X, Y) = \phi_{\text{local},2}(X, Y), \tag{7.18}$$
$$\phi_{k_3}(X, Y) = \phi_{\text{local},3}(X, Y), \tag{7.19}$$
$$\phi_j(X, Y) = 0, \qquad j \neq k_1, j \neq k_2, j \neq k_3. \tag{7.20}$$

The global basis functions defined above may be written, if desired, as functions of x and y on each element, by inverting the transformation given by Eqs. (7.12) and (7.13), giving X and Y as functions of x and y. We shall see, however, that we do not need to do this to assemble the system of algebraic equations satisfied by the finite element solution. We do, however, illustrate an example global basis function as a function of x and y in Fig. 7.6.

We now derive a useful property of the global basis functions defined by Eqs. (7.12)–(7.20). Remember that the node of the canonical triangle located at $X = 0, Y = 0$ is mapped to node k_1 of element k by Eqs. (7.12) and (7.13). We then see that, on element k,

$$
\begin{aligned}
\phi_{k_1}(x = x_{k_1}, y = y_{k_1}) &= \phi_{k_1}(X = 0, Y = 0) \\
&= \phi_{\text{local},1}(0, 0) \\
&= 1, \\
\phi_{k_2}(x = x_{k_1}, y = y_{k_1}) &= \phi_{k_2}(X = 0, Y = 0) \\
&= \phi_{\text{local},2}(0, 0) \\
&= 0, \\
\phi_{k_3}(x = x_{k_1}, y = y_{k_1}) &= \phi_{k_3}(X = 0, Y = 0) \\
&= \phi_{\text{local},3}(0, 0) \\
&= 0, \\
\phi_j(x = x_{k_1}, x = y_{k_1}) &= 0, \qquad\qquad j \neq k_1, j \neq k_2, j \neq k_3,
\end{aligned}
$$

where we have emphasised above whether we are working in the (x, y)-coordinate system, or the (X, Y)-coordinate system. Similar consideration of the nodes k_2 and k_3 allows us to write

$$
\begin{aligned}
\phi_{k_1}(x_{k_2}, y_{k_2}) &= 0, & \phi_{k_1}(x_{k_3}, y_{k_3}) &= 0, \\
\phi_{k_2}(x_{k_2}, y_{k_2}) &= 1, & \phi_{k_2}(x_{k_3}, y_{k_3}) &= 0, \\
\phi_{k_3}(x_{k_2}, y_{k_2}) &= 0, & \phi_{k_3}(x_{k_3}, y_{k_3}) &= 1, \\
\phi_j(x_{k_2}, y_{k_2}) &= 0, & \phi_j(x_{k_3}, y_{k_3}) &= 0, \qquad j \neq k_1, j \neq k_2, j \neq k_3.
\end{aligned}
$$

Hence, on any element k, the global basis functions ϕ_j, $j = 1, 2, \ldots, N_{\text{node}}$, satisfy the usual condition that

$$
\phi_j(x_i) = \begin{cases} 1, & i = j, \\ 0, & i \neq j. \end{cases} \tag{7.21}
$$

As the property given by Eq. (7.21) holds on every element, we may deduce that it holds for the whole domain.

7.4 Sets of Functions

We now define the sets of functions S^h, S^h_E, S^h_0 that will be used when specifying the finite element solution. These mirror the sets that were specified for functions of one variable in Sect. 3.3.3. In this chapter, we will calculate a finite element solution that is a linear approximation to the solution of the differential equation on each triangular element. An appropriate set S^h is

$$S^h = \left\{ \text{functions } V(x, y) \text{ given by } V(x, y) = \sum_{j=1}^{N_{\text{node}}} V_j \phi_j(x, y) \text{ where } \phi_j(x, y), \right.$$

$$\left. j = 1, 2, \ldots, N_{\text{node}}, \text{ are defined by Eqs. (7.12)–(7.20)} \right\}. \qquad (7.22)$$

Subsets of S^h, analogous to Eqs. (3.33) and (3.34), are defined as follows:

$$S_E^h = \left\{ \text{functions } V(x, y) \text{ in } S^h \text{ such that } V(x, y) \text{ satisfies all Dirichlet boundary} \right.$$

$$\left. \text{conditions} \right\}, \qquad (7.23)$$

$$S_0^h = \left\{ \text{functions } V(x, y) \text{ in } S^h \text{ such that } V(x, y) = 0 \text{ at all points } (x, y) \text{ where} \right.$$

$$\left. \text{Dirichlet boundary conditions are specified} \right\}. \qquad (7.24)$$

7.5 The Finite Element Solution

We now write the finite element solution as a linear sum of the global basis functions defined by Eqs. (7.12)–(7.20):

$$U(x, y) = \sum_{j=1}^{N_{\text{node}}} U_j \phi_j(x, y), \qquad (7.25)$$

and note that $U(x, y)$ is a member of the set S^h. On using Eqs. (7.21) and Eqs. (7.25) we see that, for $i = 1, 2, \ldots, N_{\text{node}}$,

$$\begin{aligned} U(x_i, y_i) &= \sum_{j=1}^{N_{\text{node}}} U_j \phi_j(x_i, y_i) \\ &= U_i, \end{aligned} \qquad (7.26)$$

and so U_i is, as usual, the value taken by the finite element solution at node i of the mesh.

The coefficients of the basis functions $U_1, U_2, \ldots, U_{N_{\text{node}}}$ that appear in Eq. (7.25) are determined by demanding that $U(x, y)$ satisfies the following modification of the weak formulation, Eq. (7.11):

find $U(x, y) \in S_E^h(\Omega)$ such that

$$\int_{\Omega} (\nabla \phi_i) \cdot (\nabla U) \, \mathrm{d}A = \int_{\Omega} \phi_i \, \mathrm{d}A, \tag{7.27}$$

for all $\phi_i(x, y) \in S_0^h(\Omega)$.

7.6 The System of Algebraic Equations

The finite element solution given by Eq. (7.25) contains N_{node} unknown quantities $U_1, U_2, \ldots, U_{N_{\text{node}}}$. We must therefore use the definition of the finite element solution, given by Eq. (7.27), to identify N_{node} algebraic equations to determine these values. These equations are, as usual, drawn from two sources: (i) equations that ensure that the finite element solution satisfies the Dirichlet boundary conditions; and (ii) equations that arise from substituting suitable test functions $\phi_i(x, y) \in S_0^h(\Omega)$ into the integral given by Eq. (7.27). We now discuss these two categories of equations separately.

7.6.1 Satisfying the Dirichlet Boundary Conditions

We noted earlier that $U(x, y)$, as given by Eq. (7.25), lies in the set S^h. One of the conditions imposed on the finite element solution is that it lies in the set S_E^h that is a subset of S^h. To ensure that $U(x, y)$ lies in S_E^h, we therefore need to ensure that it satisfies the Dirichlet boundary conditions. In our model problem, Dirichlet boundary conditions are applied on the whole of the boundary. Suppose node i lies on boundary, $\partial\Omega$. Using Eq. (7.26), we see that the Dirichlet boundary conditions, Eqs. (7.2)–(7.5), are satisfied provided that

$$U_i = 0, \qquad (x_i, y_i) \in \partial\Omega, \tag{7.28}$$

which provides our first category of algebraic equations.

Further, using Eq. (7.21) we see that $\phi_i(x_i, y_i) = 1$. Hence, the basis function $\phi_i(x, y)$ is not zero at all points on the boundary, where Dirichlet boundary conditions are applied. As a consequence, $\phi_i(x, y)$ is not a member of the set S_0^h and may not be used as a test function in the weak formulation given by Eq. (7.27).

7.6.2 Using Suitable Test Functions

Suppose that node i does not lie on the boundary, $\partial\Omega$. Then it follows that $\phi_i(x, y) \in S_0^h(\Omega)$. Noting that Eq. (7.25) allows us to write

$$\nabla U(x, y) = \sum_{j=1}^{N_{\text{node}}} U_j \nabla \phi_j(x, y),$$

we may substitute $\phi_i(x, y)$ into Eq. (7.27) to obtain, after a little manipulation,

$$\sum_{j=1}^{N_{\text{node}}} \left(\int_{\Omega} (\nabla \phi_i) \quad (\nabla \phi_j) \, \mathrm{d}A \right) U_j = \int_{\Omega} \phi_i \, \mathrm{d}A,$$

which may be written as

$$\sum_{j=1}^{N_{\text{node}}} A_{i,j} U_j = b_i,$$

where

$$A_{i,j} = \int_{\Omega} (\nabla \phi_i) \cdot \left(\nabla \phi_j\right) \, \mathrm{d}A, \qquad j = 1, 2, \ldots, N_{\text{node}}, \tag{7.29}$$

$$b_i = \int_{\Omega} \phi_i \, \mathrm{d}A. \tag{7.30}$$

7.6.3 The Linear System

We may now write the system of algebraic equations satisfied by $U_1, U_2, \ldots, U_{N_{\text{node}}}$ as the linear system

$$A\mathbf{U} = \mathbf{b}. \tag{7.31}$$

If node i lies on the boundary, then Eq. (7.28) applies. The entries of A and $\mathbf{b}$ in row i are then given by

$$A_{i,j} = \begin{cases} 1, & j = i, \\ 0, & j \neq i, \end{cases} \tag{7.32}$$

$$b_i = 0. \tag{7.33}$$

If node i does not lie on the boundary, then the entries of A and $\mathbf{b}$ in row i are given by Eqs. (7.29) and (7.30).

The system of algebraic equations derived above takes the same form as those derived in earlier chapters. Some entries of the matrix A and vector $\mathbf{b}$ take the form of integrals over the whole domain, while other entries may be written down explicitly.

7.7 Assembling the System of Algebraic Equations

As in earlier chapters, we will assemble the linear system given by Eq. (7.31) by first evaluating the entries of A and $\mathbf{b}$ that are given by integrals, that is, Eqs. (7.29) and (7.30). Having done this, we will then set the entries that are given explicitly by Eqs. (7.32) and (7.33). We now describe how these two categories of entries are handled in practice.

7.7.1 Assembling the Entries Defined by Integrals

We write the entries of the linear system that are defined by integrals, Eqs. (7.29) and (7.30), as the sum of integrals over individual elements: when node i does not lie on the boundary, we have

$$A_{i,j} = \sum_{k=1}^{N_{\text{ele}}} \int_{e_k} (\nabla \phi_i) \cdot \left(\nabla \phi_j\right) \, \mathrm{d}A, \qquad j = 1, 2, \ldots, N_{\text{node}}, \tag{7.34}$$

$$b_i = \sum_{k=1}^{N_{\text{ele}}} \int_{e_k} \phi_i \, \mathrm{d}A, \tag{7.35}$$

where N_{ele} is the number of elements in the mesh, and e_k is the region occupied by element k. We now explain how the integrals over individual elements may be evaluated.

Suppose the nodes of element k are the nodes k_1, k_2 and k_3, listed in an anticlockwise order. Using the definition of the global basis functions, given by Eqs. (7.12)–(7.20), we see that the only global basis functions that are not identically zero on this element are ϕ_{k_1}, ϕ_{k_2} and ϕ_{k_3}. Hence, using the definition of the entries of A given by Eq. (7.34), the only nonzero contributions from this element to the global matrix A are to the entries $A_{i,j}$ where $i = k_1, k_2, k_3$ and $j = k_1, k_2, k_3$. We therefore calculate only these nonzero local contributions, and store them in the 3×3 matrix $A^{(k)}_{\text{local}}$, where $A^{(k)}_{\text{local},i,j}$ contributes to A_{k_i,k_j} for $i = 1, 2, 3$, $j = 1, 2, 3$. Similarly, the contributions from this element to the entries of $\mathbf{b}$ given by Eq. (7.35) are zero unless $i = k_1, k_2, k_3$. We again calculate only these nonzero contributions, which we store in the local vector $\mathbf{b}^{(k)}_{\text{local}}$ that contains three entries, where $b^{(k)}_{\text{local},i}$ contributes to b_{k_i} for $i = 1, 2, 3$.

The entries of $A^{(k)}_{\text{local}}$ and $\mathbf{b}^{(k)}_{\text{local}}$ may be calculated using the transformation between element k and the canonical element given by Eqs. (7.12) and (7.13), and writing the global basis functions in terms of the local basis functions given by Eqs. (7.17)–(7.19). We derive the entries of $A^{(k)}_{\text{local}}$ first. For $i = 1, 2, 3$, $j = 1, 2, 3$, we may write

$$
\begin{aligned}
A^{(k)}_{\text{local},\,i,j} &= \int_{e_k} \left(\nabla \phi_{k_i}\right) \cdot \left(\nabla \phi_{k_j}\right) \, \mathrm{d}A \\
&= \int_{e_k} \frac{\partial \phi_{k_i}}{\partial x}\frac{\partial \phi_{k_j}}{\partial x} + \frac{\partial \phi_{k_i}}{\partial y}\frac{\partial \phi_{k_j}}{\partial y} \, \mathrm{d}A \\
&= \int_{\triangle} \left(\frac{\partial \phi_{\text{local},i}}{\partial x}\frac{\partial \phi_{\text{local},j}}{\partial x} + \frac{\partial \phi_{\text{local},i}}{\partial y}\frac{\partial \phi_{\text{local},j}}{\partial y} \right) \det(F) \, \mathrm{d}A_X, \qquad (7.36)
\end{aligned}
$$

where $\int_{\triangle}$, combined with $\mathrm{d}A_X$, denotes integration over the canonical element, and F is the Jacobian matrix of the transformation defined by Eqs. (7.12) and (7.13), which may be written as

$$
\begin{aligned}
F &= \begin{pmatrix} \frac{\partial x}{\partial X} & \frac{\partial x}{\partial Y} \\ \frac{\partial y}{\partial X} & \frac{\partial y}{\partial Y} \end{pmatrix} \\
&= \begin{pmatrix} x_{k_2} - x_{k_1} & x_{k_3} - x_{k_1} \\ y_{k_2} - y_{k_1} & y_{k_3} - y_{k_1} \end{pmatrix}. \qquad (7.37)
\end{aligned}
$$

The determinant of F, which appears in Eq. (7.36), takes a constant value on element k, given by

$$
\det(F) = \left(x_{k_2} - x_{k_1}\right)\left(y_{k_3} - y_{k_1}\right) - \left(x_{k_3} - x_{k_1}\right)\left(y_{k_2} - y_{k_1}\right). \qquad (7.38)
$$

The integrand in Eq. (7.36) contains partial derivatives of local basis functions with respect to the coordinates x and y. As the local basis functions are written in terms of the coordinates X and Y, we use the chain rule to evaluate the quantities required. The chain rule may be written in matrix form as

$$
\begin{aligned}
\begin{pmatrix} \frac{\partial \phi_{\text{local},1}}{\partial x} & \frac{\partial \phi_{\text{local},1}}{\partial y} \\ \frac{\partial \phi_{\text{local},2}}{\partial x} & \frac{\partial \phi_{\text{local},2}}{\partial y} \\ \frac{\partial \phi_{\text{local},3}}{\partial x} & \frac{\partial \phi_{\text{local},3}}{\partial y} \end{pmatrix} &= \begin{pmatrix} \frac{\partial \phi_{\text{local},1}}{\partial X} & \frac{\partial \phi_{\text{local},1}}{\partial Y} \\ \frac{\partial \phi_{\text{local},2}}{\partial X} & \frac{\partial \phi_{\text{local},2}}{\partial Y} \\ \frac{\partial \phi_{\text{local},3}}{\partial X} & \frac{\partial \phi_{\text{local},3}}{\partial Y} \end{pmatrix} \begin{pmatrix} \frac{\partial X}{\partial x} & \frac{\partial X}{\partial y} \\ \frac{\partial Y}{\partial x} & \frac{\partial Y}{\partial y} \end{pmatrix} \\
&= \begin{pmatrix} -1 & -1 \\ 1 & 0 \\ 0 & 1 \end{pmatrix} \begin{pmatrix} \frac{\partial X}{\partial x} & \frac{\partial X}{\partial y} \\ \frac{\partial Y}{\partial x} & \frac{\partial Y}{\partial y} \end{pmatrix}, \qquad (7.39)
\end{aligned}
$$

where the partial derivatives of the local basis functions with respect to X and Y follow from their definitions, Eqs. (7.14)–(7.16). It can be shown that

$$
\begin{aligned}
\begin{pmatrix} \frac{\partial X}{\partial x} & \frac{\partial X}{\partial y} \\ \frac{\partial Y}{\partial x} & \frac{\partial Y}{\partial y} \end{pmatrix} &= \begin{pmatrix} \frac{\partial x}{\partial X} & \frac{\partial x}{\partial Y} \\ \frac{\partial y}{\partial X} & \frac{\partial y}{\partial Y} \end{pmatrix}^{-1} \\
&= F^{-1},
\end{aligned}
$$

and so we may write, on using Eq. (7.37),

$$\begin{pmatrix} \frac{\partial X}{\partial x} & \frac{\partial X}{\partial y} \\ \frac{\partial Y}{\partial x} & \frac{\partial Y}{\partial y} \end{pmatrix} = \begin{pmatrix} x_{k_2} - x_{k_1} & x_{k_3} - x_{k_1} \\ y_{k_2} - y_{k_1} & y_{k_3} - y_{k_1} \end{pmatrix}^{-1}$$
$$= \frac{1}{\det(F)} \begin{pmatrix} y_{k_3} - y_{k_1} & x_{k_1} - x_{k_3} \\ y_{k_1} - y_{k_2} & x_{k_2} - x_{k_1} \end{pmatrix}, \tag{7.40}$$

where we have used an explicit expression for the inverse of a 2×2 matrix. Combining Eqs. (7.39) and (7.40) allows us to evaluate the partial derivatives that appear in Eq. (7.36). Noting that the Jacobian is given by Eq. (7.38), we now have explicit expressions for all terms in the integrand in Eq. (7.36). Finally, we see that all terms in the integrand are constant, and so the integral may be evaluated by simply multiplying this constant integrand by the area of the canonical triangle—i.e. the value 0.5—to give, for $i = 1, 2, 3$, $j = 1, 2, 3$:

$$A^{(k)}_{\text{local},\, i,j} = \frac{1}{2} \det(F) \left(\frac{\partial \phi_{\text{local},i}}{\partial x} \frac{\partial \phi_{\text{local},j}}{\partial x} + \frac{\partial \phi_{\text{local},i}}{\partial y} \frac{\partial \phi_{\text{local},j}}{\partial y} \right). \tag{7.41}$$

The entries of $\mathbf{b}^{(k)}_{\text{local}}$ may be calculated in a similar fashion. We first note that, for $i = 1, 2, 3$, integration of the local basis functions over the canonical element may be written

$$\int_{\triangle} \phi_{\text{local},i}(X, Y)\, \mathrm{d}A_X = \int_{Y=0}^{1} \left(\int_{X=0}^{1-Y} \phi_{\text{local},i}(X, Y)\, \mathrm{d}X \right) \mathrm{d}Y$$
$$= \frac{1}{6}.$$

Using the definition of $\mathbf{b}^{(k)}_{\text{local}}$ given by Eq. (7.35), and the expression for the determinant of F given by Eq. (7.38), we may write, for $i = 1, 2, 3$:

$$b^{(k)}_{\text{local},i} = \int_{e_k} \phi_{k_i}\, \mathrm{d}A$$
$$= \int_{\triangle} \phi_{\text{local},i}(X, Y) \det(F)\, \mathrm{d}A_X$$
$$= \frac{1}{6} \det(F). \tag{7.42}$$

We may now assemble the entries of A and $\mathbf{b}$ that are given by Eqs. (7.34) and (7.35) by looping over all elements, and calculating only the nonzero contributions from each element, given by Eqs. (7.41) and (7.42). These local contributions from each element are then added to the appropriate entry of the global matrix A and global vector $\mathbf{b}$; the relationship between the indices of the entries in the global linear system and the indices of the entries of local contributions is given in Table 7.1.

Table 7.1 The relationship between the location of the entries in the local and global matrices and vectors

Local index	Global index
1	$k_1 =$ `lnods(k,1)`
2	$k_2 =$ `lnods(k,2)`
3	$k_3 =$ `lnods(k,3)`

7.7.2 Setting the Entries Defined Explicitly

When looping over all elements, as described in Sect. 7.7.1, we will add contributions into the global matrix A and global vector $\mathbf{b}$ from every node of every element. This will include not only the rows where the entries of A and $\mathbf{b}$ are given by Eqs. (7.29) and (7.30), but also the rows where A and $\mathbf{b}$ are given by Eqs. (7.32) and (7.33). We therefore set these entries correctly by using the explicit values given by Eqs. (7.32) and (7.33).

7.8 Evaluating the Solution at a Given Point

Suppose we want to evaluate the finite element solution at some point (x, y), i.e. we want to evaluate $U(x, y)$ for given x and y. The first task is to identify the element that this point lies in. This may be done by first writing the transformation between element k and the canonical element, Eqs. (7.12) and (7.13), as

$$X = \frac{(x - x_{k_1})(y_{k_3} - y_{k_1}) - (y - y_{k_1})(x_{k_3} - x_{k_1})}{(x_{k_2} - x_{k_1})(y_{k_3} - y_{k_1}) - (y_{k_2} - y_{k_1})(x_{k_3} - x_{k_1})}, \tag{7.43}$$

$$Y = \frac{(x - x_{k_1})(y_{k_2} - y_{k_1}) - (y - y_{k_1})(x_{k_2} - x_{k_1})}{(x_{k_3} - x_{k_1})(y_{k_2} - y_{k_1}) - (y_{k_3} - y_{k_1})(x_{k_2} - x_{k_1})}. \tag{7.44}$$

If the point (x, y) is mapped by this transformation to a point inside the canonical triangle—that is, $X \geq 0$, $Y \geq 0$, and $X + Y \leq 1$—then (x, y) lies inside element k. Having identified the element k that contains the point (x, y), suppose that the nodes of this element are k_1, k_2, k_3. Using the definition of local basis functions given by Eqs. (7.14)–(7.16), and the values of X, Y, given by Eqs. (7.43) and (7.44), we may deduce that

$$U(x, y) = U_{k_1}\phi_{\text{local},1}(X, Y) + U_{k_2}\phi_{\text{local},2}(X, Y) + U_{k_3}\phi_{\text{local},3}(X, Y).$$

7.9 Computational Implementation

We now provide a computational implementation of the material presented in this chapter. In Sect. 7.3.1, we partitioned the domain $0 < x < 1, 0 < y < 1$ into triangular elements by first partitioning this region into equally sized rectangles, with N_x rectangles in the x-direction, and N_y rectangles in the y-direction. Listing 7.1 contains a MATLAB function that accepts N_x and N_y as input and returns the coordinates of the nodes $\mathbf{x}$, $\mathbf{y}$, the connectivity array $lnods$ and the finite element solution at each node $\mathbf{U}$.

In line 4, a function is called as `GenerateMesh`. This function generates the arrays $\mathbf{x}$, $\mathbf{y}$ and $lnods$ that are discussed in the text. It also generates a further array, `dir_nodes`, that contains a list of all nodes on the Dirichlet boundary. The body of the function, given in lines 42–80, populates the arrays $\mathbf{x}$, $\mathbf{y}$ and $lnods$ using the conventions given in Sect. 7.3.1. Note that the entries of $\mathbf{x}$ and $\mathbf{y}$ are allocated uniformly across the intervals $0 \leq x \leq 1$ and $0 \leq y \leq 1$—these lines of code need to be modified if a different size of rectangular domain is used. The entries of the array `dir_nodes` are ordered starting from the point $x = 0$, $y = 0$ and proceeding around the boundary in an anticlockwise direction. The ordering of the nodes in the array `dir_nodes`, however, has no effect on the finite element solution calculated.

For a mesh with N_x elements in the x-direction and N_y elements in the y-direction, there will be a total of $(N_x + 1)(N_y + 1)$ nodes in the mesh, and the linear system will therefore contain $(N_x + 1)(N_y + 1)$ equations. In lines 7–8, we initialise the global matrix A and global vector $\mathbf{b}$ to be of the correct size, with all entries equal to zero. Note that we define A to be a sparse matrix—this has the effect that only the nonzero entries of A are stored. As each row of A contains very few nonzero entries, this results in a significant reduction in computational memory requirement, particularly when the number of rows in the matrix is large. A discussion of sparse matrices is given in Appendix A.

There will be $2N_xN_y$ elements in the mesh defined in Sect. 7.3.1. In lines 12–19, we loop over these elements, calculating the local contributions to A and $\mathbf{b}$, before incrementing the appropriate entries of A and $\mathbf{b}$. The local entries are calculated using the function `CalculateContributionOnElement` that occupies lines 85–117. In this function, the location of the nodes of the element is given in lines 89–94. The determinant of the transformation between the element and the canonical element is given in line 98, before the partial derivatives of the basis functions, with respect to first the (X, Y)-coordinate system and then the (x, y)-coordinate system, are calculated in lines 102–104, using the expression given by Eq. (7.39). These partial derivatives are used to calculate the local contribution to A (given by Eq. (7.41)) in lines 107–114, before assigning the local contributions to $\mathbf{b}$ (given by Eq. (7.42)) in line 117.

Having calculated the contributions to A and $\mathbf{b}$ that arise from integrals, we then set entries that enforce the Dirichlet boundary conditions in lines 22–26. Note that we set to zero all entries of the rows of A that correspond to Dirichlet boundary conditions in line 22 as some of these entries may have been set to nonzero values when looping over all elements—see Sect. 7.7.2 for more details. We then solve the linear system. In Appendix A, we propose solving large linear systems using preconditioned, iterative methods. Lines 29–32 solve the linear system using an incomplete LU-factorisation (calculated in line 29), preconditioned GMRES (line 32). When using GMRES, we use a restart value of 30. For more details on what these terms mean, see Appendix A. Finally, the solution is plotted using lines 35–38.

This MATLAB function may be executed by typing, for example,

```
[x, y, lnods, U] = Chap7_CalculateExampleFem(50, 50);
```

into a MATLAB session.

Listing 7.1 `Chap7_CalculateExampleFem.m`, a MATLAB function for calculating the finite element solution of the model problem.

```
1  function [x, y, lnods, U] = Chap7_CalculateExampleFem(Nx,Ny)
2
3  % Generate data structures needed
4  [x, y, lnods, dir_nodes] = GenerateMesh(Nx, Ny);
5
6  % Initialise A and b to zero
7  A = sparse((Nx+1)*(Ny+1), (Nx+1)*(Ny+1));
8  b = zeros((Nx+1)*(Ny+1), 1);
9
10 % Loop over elements, calculating local contributions, and
11 % adding the local contributions into the global A and b
12 for k=1:2*Nx*Ny
13     [A_local, b_local] = ...
14         CalculateContributionOnElement(x, y, lnods, k);
15
16     A(lnods(k,:), lnods(k,:)) = ...
17         A(lnods(k,:), lnods(k,:)) + A_local;
18     b(lnods(k,:)) = b(lnods(k,:)) + b_local;
19 end
20
21 % Handle Dirichlet boundary conditions
22 A(dir_nodes,:) = 0;
23 for i=1:length(dir_nodes)
24     A(dir_nodes(i),dir_nodes(i)) = 1;
25 end
26 b(dir_nodes) = 0;
27
28 % Compute ILU factorization of A
29 [Lower, Upper] = ilu(A);
30
31 % Use ILU preconditioned GMRES to solve linear system
32 U = gmres(A, b, 30, [], 1000, Lower, Upper);
33
34 % Plot solution
35 trisurf(lnods, x, y, U);
36 xlabel('x')
37 ylabel('y')
38 zlabel('U')
39
40
41 % Function to generate data structures required by the mesh
42 function [x, y, lnods, dir_nodes] = GenerateMesh(Nx, Ny)
43
```

```
% Set entries of lnods
lnods = zeros(2*Nx*Ny, 3);
for j=1:Ny;
    for i=1:Nx;
        ele = 2*(j-1)*Nx+2*i-1;
        lnods(ele,:) = ...
            [(j-1)*(Nx+1)+i, (j-1)*(Nx+1)+i+1, j*(Nx+1)+i];

        ele = ele+1;
        lnods(ele,:) = ...
            [j*(Nx+1)+i+1, j*(Nx+1)+i, (j-1)*(Nx+1)+i+1];
    end
end

% Set entries of x and y over the interval from 0 to 1
x = zeros((Nx+1)*(Ny+1),1);
y = zeros((Nx+1)*(Ny+1),1);

for i=1:Nx+1
    x(i:Nx+1:(Nx+1)*(Ny+1)) = (i-1)/Nx;
end

for j=1:Ny+1
    y((j-1)*(Nx+1)+1:j*(Nx+1)) = (j-1)/Ny;
end

% Specify locations of Dirichlet boundary conditions
% Nodes on y=0
dir_nodes(1:Nx+1) = 1:Nx+1;
% Nodes on x=1
dir_nodes(Nx+2:Nx+Ny+1) = 2*(Nx+1):Nx+1:(Ny+1)*(Nx+1);
% Nodes on y=1
dir_nodes(Nx+Ny+2:2*Nx+Ny+1) = ...
     (Ny+1)*(Nx+1)-1:-1:Ny*(Nx+1)+1;
% Nodes on x=0
dir_nodes(2*Nx+Ny+2:2*(Nx+Ny)) = ...
    (Ny-1)*(Nx+1)+1:-(Nx+1):Nx+2;



% Calculate local contributions to A and b from element k
function [A_local, b_local] = ...
    CalculateContributionOnElement(x, y, lnods, k)

% Define x and y coordinate of each node
x1 = x(lnods(k,1));
x2 = x(lnods(k,2));
x3 = x(lnods(k,3));
y1 = y(lnods(k,1));
y2 = y(lnods(k,2));
y3 = y(lnods(k,3));

% Calculate Jacobian of transformation between canonical
% element and element k
detF = (x2-x1)*(y3-y1) - (x3-x1)*(y2-y1);

% Evaluate derivatives of basis functions in terms of (x,y)
% coordinates
Finv = [y3-y1, x1-x3; y1-y2, x2-x1]/detF;
```

```
dphi_dX = [-1, -1; 1, 0; 0, 1];
dphi_dx = dphi_dX*Finv;

% Assemble local contribution to A
A_local = zeros(3,3);
for i=1:3
    for j=1:3
        A_local(i,j) = detF*0.5* ...
            (dphi_dx(i,1)*dphi_dx(j,1) + ...
            dphi_dx(i,2)*dphi_dx(j,2));
    end
end

% Assemble local contribution to b
b_local = ones(3,1)*detF/6;
```

In the computational implementation presented in this section, we chose to use a sparse matrix to store the matrix A and to use a preconditioned iterative technique to solve the linear system that arises. For this simple example problem, it is likely that a sufficiently accurate finite element solution may be obtained by setting both N_x and N_y to small enough values that A may be stored as a full matrix. The linear system may then be solved using a direct solver, such as the backslash operator in MATLAB. For more complex partial differential equations, the linear systems may be larger, and using sparse matrices and preconditioned iterative techniques will be the only practical option for storing and solving these systems. We therefore feel that the time is right to introduce the linear algebra techniques used in this implementation.

7.10 More Complex Geometries

Calculation of the finite element solution of a differential equation requires us to partition the domain on which the solution is defined into elements. In this chapter, we considered only the square domain $0 < x < 1$, $0 < y < 1$, which could be partitioned manually. Partitioning other geometries into triangular elements, even the simple circular domain $x^2 + y^2 < 1$ for example, is not so straightforward. Furthermore, problems of practical interest are often posed on domains that are irregular. Fortunately, for two-dimensional problems, two excellent, freely available packages are available for partitioning a subset Ω of the (x, y)-plane into a mesh of triangular elements. One of these is `DistMesh` [7], which may be interfaced from MATLAB. The other is `triangle` [8,9], written in the programming language C. Both packages can easily be found using a search engine, and documentation is available explaining how they may be used to generate a mesh of triangular elements for a given domain Ω.

We now give some general comments on using mesh generation packages that are not aimed at any specific package. A mesh generation package will replace the function `GenerateMesh` called from line 4 of Listing 7.1. This function returned the coordinates of the nodes $\mathbf{x}$, $\mathbf{y}$, the connectivity array $lnods$ and the list of nodes that are on the part of the boundary where Dirichlet boundary conditions are applied. In general, mesh generation packages may be used to create at least the first three of these. One word of caution is that the user should perform a check to ensure that the connectivity array does list the nodes of each element in an anticlockwise direction. This is straightforward to do—if the determinant of the Jacobian of the transformation between an element and the canonical element, given by Eq. (7.38), is negative, then the nodes are not listed in the correct order. If the nodes are not ordered correctly, then this may be fixed by interchanging the order of two nodes of that element in the connectivity array. Suppose a list of nodes lying on the boundary where Dirichlet boundary conditions are applied is not available. This list may easily be generated—simply loop over all nodes, and check whether they lie on this boundary.

7.11 Exercises

7.1 Modify the computational implementation given in Listing 7.1 so that it may be used to solve the partial differential equation

$$-\left(\frac{\partial^2 u}{\partial x^2}+\frac{\partial^2 u}{\partial y^2}\right)=5, \qquad -3<x<3, \quad -1<y<0,$$

subject to Dirichlet boundary conditions on the whole boundary, given by

$$\begin{aligned} u(x,-1) &= (x+3)(x-3), & -3<x<3,\\ u(3,y) &= 0, & -1<y<0,\\ u(x,0) &= 0, & -3<x<3,\\ u(-3,y) &= 0, & -1<y<0. \end{aligned}$$

Use the material presented in Sect. 7.8 to evaluate $U(0.48,-0.2)$.

7.2 Use a mesh generation package such as `DistMesh`, described in Sect. 7.10, to generate a mesh of triangular elements that is a partitioning of the square $0<x<1$, $0<y<1$. Modify Listing 7.1 so that it uses the mesh that you have generated. Confirm that your implementation is correct by comparison with the output from the original implementation given by Listing 7.1.

7.3 In this exercise, we will calculate the finite element solution of the differential equation given by

$$-\left(\frac{\partial^2 u}{\partial x^2}+\frac{\partial^2 u}{\partial y^2}\right)=1, \qquad x^2+y^2<1,$$

subject to Dirichlet boundary conditions on the whole boundary, given by

$$u(x, y) = 0, \qquad x^2 + y^2 = 1.$$

This boundary value problem has solution

$$u(x, y) = \frac{1}{4}\left(1 - x^2 - y^2\right).$$

(a). Use a mesh generation package such as `DistMesh`, described in Sect. 7.10, to generate a mesh of triangular elements that is a partitioning of the domain $x^2 + y^2 = 1$.
(b). Modify Listing 7.1 so that it uses the mesh generated in part (a) to calculate the finite element solution of the boundary value problem defined above. Verify your solution is correct by comparison with the analytic solution.

7.4 Modify Listing 7.1 so that (i) the matrix A is stored as a full matrix, rather than a sparse matrix; and (ii) the linear system is solved using the MATLAB's backslash operator. Investigate the performance of both implementations by varying the values of N_x and N_y used as input.

Chapter 8
More General Elliptic Problems

In the previous chapter, we explained how to calculate the finite element solution of a very simple elliptic partial differential equation, where Dirichlet boundary conditions were applied on the whole boundary. We develop these ideas in this chapter to allow the application of the finite element method to more general linear, elliptic partial differential equations, and more general boundary conditions.

8.1 A Model Problem

In this chapter, we will illustrate the application of the finite element method to the model partial differential equation

$$-\nabla \cdot (p(x, y)\nabla u) + \mathbf{q}(x, y) \cdot \nabla u + r(x, y)u = f(x, y), \qquad (x, y) \in \Omega, \tag{8.1}$$

where Ω is some finite region of the (x, y)-plane, the vector-valued function $\mathbf{q}(x, y)$ is given by

$$\mathbf{q}(x, y) = \begin{pmatrix} q_1(x, y) \\ q_2(x, y) \end{pmatrix},$$

and $p(x, y)$, $\mathbf{q}(x, y)$, $r(x, y)$ and $f(x, y)$ are given functions. We partition the boundary $\partial\Omega$ into two regions, $\partial\Omega_D$ and $\partial\Omega_N$. On the boundary $\partial\Omega_D$, we apply Dirichlet boundary conditions:

$$u(x, y) = g_D(x, y), \qquad (x, y) \in \partial\Omega_D, \tag{8.2}$$

and on the boundary $\partial\Omega_N$, we apply natural Neumann boundary conditions:

J. Whiteley, *Finite Element Methods*, Mathematical Engineering,
DOI 10.1007/978-3-319-49971-0_8

$$p(x, y)\,(\nabla u)\cdot \mathbf{n} = g_N(x, y), \qquad (x, y) \in \partial\Omega_N, \tag{8.3}$$

where $\mathbf{n}$ is the normal vector to the boundary $\partial\Omega_N$, pointing out of the domain Ω, and of unit length.

8.2 The Weak Formulation

The Sobolev space $H^1(\Omega)$ and the subspaces $H_E^1(\Omega)$ and $H_0^1(\Omega)$ of $H^1(\Omega)$ are defined in Sect. 7.2.1. The first step when deriving the weak formulation of the differential equation given by Eq. (8.1), subject to the boundary conditions given by Eqs. (8.2) and (8.3), is to multiply the differential equation by a test function $v(x, y) \in H_0^1(\Omega)$, and to integrate the resulting product over the domain Ω:

$$\int_\Omega (-\nabla\cdot(p\nabla u) + \mathbf{q}\cdot\nabla u + ru)\,v\,\mathrm{d}A = \int_\Omega fv\,\mathrm{d}A. \tag{8.4}$$

Using Eq. (B.2) from Appendix B allows us to write

$$-v\nabla\cdot(p\nabla u) = -\nabla\cdot(vp\nabla u) + p\,(\nabla v)\cdot(\nabla u),$$

and so Eq. (8.4) may be written

$$\int_\Omega -\nabla\cdot(vp\nabla u) + p\,(\nabla v)\cdot(\nabla u) + v\mathbf{q}\cdot\nabla u + ruv\,\mathrm{d}A = \int_\Omega fv\,\mathrm{d}A.$$

Applying the divergence theorem to the first term on the left-hand side of this equation gives

$$-\int_{\partial\Omega} vp\,(\nabla u)\cdot\mathbf{n}\,\mathrm{d}s + \int_\Omega p\,(\nabla v)\cdot(\nabla u) + v\mathbf{q}\cdot\nabla u + ruv\,\mathrm{d}A = \int_\Omega fv\,\mathrm{d}A. \tag{8.5}$$

The boundary $\partial\Omega$ was partitioned into the components $\partial\Omega_D$ and $\partial\Omega_N$. As $v(x, y) \in H_0^1(\Omega)$, we have $v(x, y) = 0$ for all points on $\partial\Omega_D$. Using this observation, and the Neumann boundary condition given by Eq. (8.3), we may write the line integral above as

$$\begin{aligned}\int_{\partial\Omega} vp\,(\nabla u)\cdot\mathbf{n}\,\mathrm{d}s &= \int_{\partial\Omega_D} vp\,(\nabla u)\cdot\mathbf{n}\,\mathrm{d}s + \int_{\partial\Omega_N} vp\,(\nabla u)\cdot\mathbf{n}\,\mathrm{d}s \\ &= 0 + \int_{\partial\Omega_N} vg_N\,\mathrm{d}s.\end{aligned}$$

Substituting this into Eq. (8.5), we obtain, after a little rearranging,

$$\int_{\Omega} p\left(\nabla v\right)\cdot\left(\nabla u\right)+v\mathbf{q}\cdot\nabla u+ruv\,\mathrm{d}A=\int_{\Omega} fv\,\mathrm{d}A+\int_{\partial\Omega_N} vg_N\,\mathrm{d}s.$$

All that is now required to state the weak formulation of the boundary value problem defined by Eqs. (8.1)–(8.3) is to demand that the weak solution $u(x, y)$ satisfies the Dirichlet boundary conditions given by Eq. (8.2). We then define the weak solution by:

find $u(x, y) \in H^1_E(\Omega)$, such that

$$\int_{\Omega} p\left(\nabla v\right)\cdot\left(\nabla u\right)+v\mathbf{q}\cdot\nabla u+ruv\,\mathrm{d}A=\int_{\Omega} fv\,\mathrm{d}A+\int_{\partial\Omega_N} vg_N\,\mathrm{d}s, \qquad (8.6)$$

for all $v(x, y) \in H^1_0(\Omega)$.

8.3 The Mesh and Basis Functions

We partition the domain Ω into a mesh of triangular elements, as described in Sect. 7.3.1. We use basis functions that allow a linear approximation to be made to the solution on each element. Suitable basis functions are the basis functions defined in Sect. 7.3.2 by Eqs. (7.12)–(7.20). The set of functions S^h, and the subsets S^h_E and S^h_0 of S^h, are then those defined in Sect. 7.4.

8.4 The Finite Element Solution

Having defined the weak solution by Eq. (8.6), the sets S^h_E and S^h_0 allow us to define the finite element solution by

find $U(x, y) \in S^h_E$, such that

$$\int_{\Omega} p\left(\nabla \phi_i\right)\cdot\left(\nabla U\right)+\phi_i\mathbf{q}\cdot\nabla U+rU\phi_i\,\mathrm{d}A=\int_{\Omega} f\phi_i\,\mathrm{d}A+\int_{\partial\Omega_N} \phi_i g_N\,\mathrm{d}s, \qquad (8.7)$$

for all $\phi_i(x, y) \in S^h_0$.

8.5 The Algebraic Equations

We write the finite element solution as:

$$U(x, y) = \sum_{j=1}^{N_{\text{node}}} U_j\phi_j(x, y), \qquad (8.8)$$

where N_{node} is the number of nodes in the mesh, and note that $U(x, y)$ written in this form is a member of the set S^h. We will assemble a system of N_{node} algebraic equations for the unknown values $U_1, U_2, \ldots, U_{N_{\text{node}}}$ using a similar procedure to that used in Sect. 7.6 for the simple model partial differential equation. As in earlier chapters, this system comprises two categories of equations. The first category of equations ensure that $U(x, y)$ satisfies the Dirichlet boundary conditions, so that $U(x, y)$ lies in the subset S_E^h of the set S^h, as required. The second category of equations arises from substituting a suitable test function $\phi_i(x)$ into Eq. (8.7). We now discuss both categories of equations separately.

8.5.1 Satisfying the Dirichlet Boundary Conditions

Suppose that node i of the mesh lies on $\partial\Omega_D$, the part of the boundary where Dirichlet boundary conditions are applied. The Dirichlet boundary condition is satisfied at this node if

$$U(x_i, y_i) = g_D(x_i, y_i).$$

Recall from Eq. (7.26) that $U(x_i, y_i) = U_i$. A suitable algebraic equation that ensures that $U(x, y)$ satisfies a Dirichlet boundary condition at node i is then

$$U_i = g_D(x_i, y_i). \tag{8.9}$$

8.5.2 Using Suitable Test Functions

Suppose that node i of the mesh does not lie on the part of the boundary where Dirichlet boundary conditions are applied. The basis function $\phi_i(x, y)$ will be nonzero at node i and will take the value zero at all other nodes. As such, $\phi_i(x, y)$ will take the value zero at all nodes on the part of the boundary where Dirichlet boundary conditions are applied, and so $\phi_i(x, y) \in S_0^h$. Noting that Eq. (8.8) allows us to write

$$\nabla U(x, y) = \sum_{j=1}^{N_{\text{node}}} U_j \nabla\phi_j(x, y),$$

we then substitute $\phi_i(x, y)$ into Eq. (8.7) and obtain, after some manipulation,

$$\sum_{j=1}^{N_{\text{node}}} \left(\int_{\Omega} p\,(\nabla\phi_i)\cdot(\nabla\phi_j) + \phi_i\mathbf{q}\cdot(\nabla\phi_j) + r\phi_i\phi_j \;\mathrm{d}A \right) U_j = \int_{\Omega} f\phi_i \;\mathrm{d}A + \int_{\partial\Omega_N} \phi_i g_N \;\mathrm{d}s. \tag{8.10}$$

8.5.3 The Linear System

We may now combine Eqs. (8.9) and (8.10) into the linear system

$$AU = \mathbf{b}, \tag{8.11}$$

where, if node i lies on $\partial\Omega_D$, we deduce from Eq. (8.9) that

$$A_{i,j} = \begin{cases} 1, & j = i, \\ 0, & j \neq i, \end{cases} \tag{8.12}$$

$$b_i = g_D(x_i, y_i), \tag{8.13}$$

and, if node i does not lie on $\partial\Omega_D$, we deduce from Eq. (8.10) that

$$A_{i,j} = \int_\Omega p\,(\nabla\phi_i)\cdot(\nabla\phi_j) + \phi_i\mathbf{q}\cdot(\nabla\phi_j) + r\phi_i\phi_j\,\mathrm{d}A, \qquad j = 1, 2, \ldots, N_{\text{node}}, \tag{8.14}$$

$$b_i = \int_\Omega f\phi_i\,\mathrm{d}A + \int_{\partial\Omega_N} \phi_i g_N\,\mathrm{d}s. \tag{8.15}$$

8.6 Assembling the Algebraic Equations

The only boundary conditions satisfied by the model partial differential equation in Chap. 7 were Dirichlet boundary conditions. The model differential equation used in this chapter contains the additional feature of some Neumann boundary conditions. This results in an extra term in some entries of the vector on the right-hand side of the linear system given by Eq. (8.11), specifically the integral over $\partial\Omega_N$ that appears in Eq. (8.15). We will see later that it is convenient to write this extra integral as the sum of contributions from individual element edges.

As in earlier chapters, the linear system given by Eq. (8.11) is assembled by first evaluating the entries given by Eqs. (8.14) and (8.15), i.e. the entries that are given by integrals over the whole domain and, for the vector $\mathbf{b}$, an extra integral over $\partial\Omega_N$. As usual, the integrals over Ω will be calculated by decomposing the integral into the sum of integrals over individual elements. The integrals over $\partial\Omega_N$ will, as explained above, be calculated by decomposing these integrals into the sum of integrals over individual element edges. We then write Eqs. (8.14) and (8.15), the equations that apply when node i does not lie on $\partial\Omega_D$, as:

$$A_{i,j} = \sum_{k=1}^{N_{\text{ele}}} \int_{e_k} p\,(\nabla\phi_i)\cdot(\nabla\phi_j) + \phi_i\mathbf{q}\cdot(\nabla\phi_j) + r\phi_i\phi_j\,\mathrm{d}A, \qquad j = 1, 2, \ldots, N_{\text{node}}, \tag{8.16}$$

$$b_i = \sum_{k=1}^{N_{\text{ele}}} \int_{e_k} f\phi_i \, \mathrm{d}A + \sum_{k=1}^{N_{\text{Neu}}} \int_{\partial e_{N_k}} \phi_i g_N \, \mathrm{d}s, \tag{8.17}$$

where N_{ele} is the number of elements in the mesh, e_k is the region occupied by element k, N_{Neu} is the number of element edges that lie on $\partial\Omega_N$ and ∂e_{N_k} is edge k of the element edges that lie on $\partial\Omega_N$.

After the entries of A and $\mathbf{b}$ given by Eqs. (8.14) and (8.15) have been assembled, we set the entries that are given explicitly by Eqs. (8.12) and (8.13). We now give more details on how these tasks may be carried out.

8.6.1 Evaluating Integrals Over Ω

We begin by calculating the contributions to Eqs. (8.16) and (8.17) that arise from integrating over an element. Suppose that element k contains the nodes k_1, k_2, k_3, listed in an anti-clockwise order. We explained in Sect. 7.7.1 that only the basis functions $\phi_{k_1}(x)$, $\phi_{k_2}(x)$, $\phi_{k_3}(x)$ are nonzero on this element, and that the nonzero contributions to Eqs. (8.16) and (8.17) from integrating over this element can be stored in a 3×3 matrix $A^{(k)}_{\text{local}}$ with entries given by, for $i = 1, 2, 3$, $j = 1, 2, 3$:

$$A^{(k)}_{\text{local},i,j} = \int_{e_k} p(x, y) \left(\nabla\phi_{k_i}\right) \cdot \left(\nabla\phi_{k_j}\right) + \phi_{k_i}\mathbf{q}(x, y) \cdot \left(\nabla\phi_{k_j}\right) + r(x, y)\phi_{k_i}\phi_{k_j} \, \mathrm{d}A,$$

and a vector with three entries $\mathbf{b}^{(k)}_{\text{local}}$, where the entries are given by, for $i = 1, 2, 3$:

$$b^{(k)}_{\text{local},i} = \int_{e_k} f(x, y)\phi_{k_i} \, \mathrm{d}A.$$

As explained in Sect. 7.7.1, the entry $A^{(k)}_{\text{local},i,j}$ contributes to entry A_{k_i,k_j} of the global matrix A for $i = 1, 2, 3$, and $j = 1, 2, 3$. Similarly, the entry $\mathbf{b}^{(k)}_{\text{local},i}$ contributes to entry $\mathbf{b}_{k_i}$ of the global vector $\mathbf{b}$ for $i = 1, 2, 3$.

We use the local basis functions given by Eqs. (7.12)–(7.20) to write these local contributions as integrals over the canonical triangle. The entries of $A^{(k)}_{\text{local},i,j}$ are given by, for $i = 1, 2, 3$, $j = 1, 2, 3$:

$$A^{(k)}_{\text{local},i,j} = \int_{\triangle} \Big(p(x, y) \left(\nabla\phi_{\text{local},i}\right) \cdot \left(\nabla\phi_{\text{local},j}\right) + \phi_{\text{local},i}\mathbf{q}(x, y) \cdot \left(\nabla\phi_{\text{local},j}\right) + \\ r(x, y)\phi_{\text{local},i}\phi_{\text{local},j} \Big) \det(F) \, \mathrm{d}A_X, \tag{8.18}$$

where, as in Chap. 7, $\int_\triangle$, coupled with $\mathrm{d}A_X$, denotes integration over the canonical triangular element and $\det(F)$ is the Jacobian of the transformation between element k and the canonical element, given by Eq. (7.37). The partial derivatives that appear in the integrand above may all be calculated using the method described in Sect. 7.7.1. To evaluate the integral above, we need to write the functions p, $\mathbf{q}$ and r as functions of X and Y, as these are the variables that we are integrating with respect to. This is done by using Eqs. (7.12) and (7.13) to evaluate x and y as functions of X and Y—we note that these equations may be written in the form:

$$x = x_{k_1}\phi_{\text{local},1}(X,Y) + x_{k_2}\phi_{\text{local},2}(X,Y) + x_{k_3}\phi_{\text{local},3}(X,Y),$$
$$y = y_{k_1}\phi_{\text{local},1}(X,Y) + y_{k_2}\phi_{\text{local},2}(X,Y) + y_{k_3}\phi_{\text{local},3}(X,Y).$$

The function $p(x, y)$, for example, may be evaluated for any values of X and Y by first using the expressions above to calculate the required values of x and y.

Similarly, the entries, of $\mathbf{b}^{(k)}_{\text{local},i,j}$ are given by, for $i = 1, 2, 3$,

$$b^{(k)}_{\text{local},i} = \int_\triangle f(x,y)\phi_{\text{local},i}\det(F)\,\mathrm{d}A_X, \tag{8.19}$$

where the function $f(x, y)$ may be written as a function of X and Y using the same method as for functions of x and y that appear in Eq. (8.18).

The local contributions to A and $\mathbf{b}$, given by Eqs. (8.18) and (8.19), are now written as integrals of functions of X and Y over the canonical element. Although we can write down explicit expressions for all terms that appear in the integrand of these equations, performing the integration analytically may be very difficult, or even impossible. For ordinary differential equations, we saw that the (one-dimensional) integrals arising may be approximated numerically using quadrature. We will explain in Sect. 8.7 how to approximate integrals over the canonical triangle, such as those given by Eqs. (8.18) and (8.19), using quadrature.

Having evaluated the local contributions to A and $\mathbf{b}$, we then add these local contributions into the appropriate entries of A and $\mathbf{b}$ as described in Sect. 7.7.1. We perform this operation on every element to assemble the integral over Ω.

8.6.2 Evaluating Integrals Over $\partial\Omega_N$

We now turn our attention to the second integral on the right-hand side of Eq. (8.17), the integral over the part of the boundary where Neumann boundary conditions are applied. Suppose that edge k of the element edges that lie on $\partial\Omega_N$, denoted by ∂e_{N_k}, connects node k_1 to node k_2. Only the basis functions ϕ_{k_1} and ϕ_{k_2} will be nonzero on this edge, and so there will be a nonzero contribution from this edge only to the entries b_{k_1} and b_{k_2} of the global vector $\mathbf{b}$. In the spirit of calculating only nonzero local contributions, we store these two nonzero local contributions to $\mathbf{b}$ from this element edge in the vector $\mathbf{b}^{(k)}_{\text{local edge}}$ with two entries given by, for $i = 1, 2$,

$$b^{(k)}_{\text{local edge},i} = \int_{\partial e_{N_k}} \phi_{k_i}(x, y) g_N(x, y)\, \mathrm{d}s. \tag{8.20}$$

As the edge ∂e_{N_k} connects node k_1 to node k_2, points (x, y) that lie on this edge are parameterised by, for $0 \le X \le 1$,

$$x = (1 - X)x_{k_1} + X x_{k_2}, \tag{8.21}$$

$$y = (1 - X)y_{k_1} + X y_{k_2}. \tag{8.22}$$

Further, we may write the basis functions ϕ_{k_1} and ϕ_{k_2} in terms of the variable X on this edge as, for $0 \le X \le 1$,

$$\phi_{k_1}(X) = 1 - X,$$

$$\phi_{k_2}(X) = X.$$

It is straightforward to justify these expressions for ϕ_{k_1} and ϕ_{k_2}—using this definition it can be seen that ϕ_{k_1} and ϕ_{k_2} are linear functions, and that they take the correct values at the node k_1 (where $X = 0$) and the node k_2 (where $X = 1$). The integral given by Eq. (8.20) may then be written, for $i = 1, 2$:

$$b^{(k)}_{\text{local edge},i} = h \int_0^1 \phi_{k_i}(X) g_N(x, y)\, \mathrm{d}X, \tag{8.23}$$

where h is the length of the edge, given by

$$h = \sqrt{\left(x_{k_1} - x_{k_2}\right)^2 + \left(y_{k_1} - y_{k_2}\right)^2},$$

and x and y are given as functions of X by Eqs. (8.21) and (8.22), allowing $g_N(x, y)$ to be evaluated along this edge as a function of X.

The integral given by Eq. (8.23) is an integral over the interval $0 \le X \le 1$. It may therefore be approximated using quadrature techniques, as described in Sect. 3.5. On each element edge, having calculated the nonzero local contributions, we then add these contributions into the global vector $\mathbf{b}$: entry $b^{(k)}_{\text{local edge},i}$ contributes to the entry b_{k_i} of $\mathbf{b}$ for $i = 1, 2$.

8.6.3 Setting Entries Defined Explicitly

Having assembled all entries of the linear system that are of the form given by Eqs. (8.14) and (8.15), we then complete the assembly of the linear system by setting the entries given by Eqs. (8.12) and (8.13). As explained in earlier chapters, it is important to explicitly set all entries given by Eqs. (8.12) and (8.13), as incorrect values may have been inserted into these entries while assembling the contributions to Eqs. (8.14) and (8.15).

8.7 Quadrature Over Triangles

Assembling the linear system, as described in Sect. 8.6, requires us to evaluate integrals of the form

$$\int_{\triangle} h(X,Y)\,\mathrm{d}A_X,$$

over the canonical element, i.e. the region occupying $X \geq 0, Y \geq 0$ and $1 - X - Y \geq 0$. We evaluated some integrals of this form in Chap. 7, but these were integrals of very simple functions that could be evaluated analytically. For general partial differential equations defined by Eqs. (8.1)–(8.3), the functions $p(x,y)$, $\mathbf{q}(x,y)$, $r(x,y)$ and $f(x,y)$ may be such that the integrals given by Eqs. (8.14) and (8.15) are very difficult, or even impossible, to evaluate analytically. In these circumstances, we require a numerical quadrature rule that allows us to approximate integrals over triangular elements, similar to that described for functions of one variable in Sect. 3.5. We now derive a suitable quadrature rule by using a map between the unit square and the canonical triangle.

In $(\hat{X},\hat{Y})$-coordinates, the unit square is given by $0 \leq \hat{X} \leq 1$, $0 \leq \hat{Y} \leq 1$. We may map between the unit square in $(\hat{X},\hat{Y})$-coordinates and the canonical triangle in (X,Y)-coordinates using the transformation defined by

$$X = \hat{X},$$
$$Y = (1-\hat{X})\hat{Y}.$$

Using this transformation, an integral over the canonical triangle of a function written in (X,Y)-coordinates is related to an integral over the unit square in $(\hat{X},\hat{Y})$-coordinates by

$$\int_{\triangle} h(X,Y)\,\mathrm{d}A_X = \int_{\hat{Y}=0}^{1}\left(\int_{\hat{X}=0}^{1} h(X,Y)(1-\hat{X})\,\mathrm{d}\hat{X}\right)\mathrm{d}\hat{Y}.$$

We write this equation as

$$\int_{\triangle} h(X,Y)\,\mathrm{d}A_X = \int_{\hat{Y}=0}^{1} \hat{h}(\hat{Y})\,\mathrm{d}\hat{Y}, \tag{8.24}$$

where

$$\hat{h}(\hat{Y}) = \int_{\hat{X}=0}^{1} h(X, Y)(1 - \hat{X})\, \mathrm{d}\hat{X}$$
$$= \int_{\hat{X}=0}^{1} h\left(\hat{X}, (1 - \hat{X})\hat{Y}\right)(1 - \hat{X})\, \mathrm{d}\hat{X}.$$

This integral is over the interval $0 \leq \hat{X} \leq 1$ and may be approximated by Gaussian quadrature as described in Sect. 3.5. Using M quadrature points, weights $w_1, w_2, \ldots, w_M$, and quadrature points $Z_1, Z_2, \ldots, Z_M$, we may write this approximation as

$$\hat{h}(\hat{Y}) = \sum_{m=1}^{M} w_m h\left(Z_m, (1 - Z_m)\hat{Y}\right)(1 - Z_m). \tag{8.25}$$

Similarly, the integral on the right-hand side of Eq. (8.24) is an integral over the interval $0 \leq \hat{Y} \leq 1$, and may be approximated, using the same Gaussian quadrature scheme, by

$$\int_{\triangle} h(X, Y)\, \mathrm{d}A_X = \sum_{n=1}^{M} w_n \hat{h}(Z_n). \tag{8.26}$$

Substituting Eqs. (8.25) into (8.26) yields our desired quadrature rule:

$$\int_{\triangle} h(X, Y)\, \mathrm{d}A_X = \sum_{m=1}^{M}\sum_{n=1}^{M} w_m w_n (1 - Z_m) h\left(Z_m, (1 - Z_m)Z_n\right). \tag{8.27}$$

8.8 Computational Implementation

We calculate the finite element solution of the model problem defined, for $-1 < x < 1, -1 < y < 1$, by

$$-\nabla \cdot \left(\mathrm{e}^x \nabla u\right) + 3x\frac{\partial u}{\partial x} + 2y\frac{\partial u}{\partial y} - 6u = -2\mathrm{e}^x\left(1 + x + 6y\right),$$

subject to Dirichlet boundary conditions

$$u(x, -1) = x^2 - 2, \qquad -1 < x < 1,$$
$$u(1, y) = 2y^3 + 1, \qquad -1 < y < 1,$$

and natural Neumann boundary conditions

$$e^x (\nabla u) \cdot \mathbf{n} = 6e^x, \qquad -1 < x < 1, \qquad y = 1,$$
$$e^x (\nabla u) \cdot \mathbf{n} = 2e^x, \qquad x = -1, \qquad -1 < y < 1.$$

This boundary value problem has solution $u(x, y) = x^2 + 2y^3$.

A computational implementation for calculating the solution of the model problem above, using the material presented in this chapter, is given in Listing 8.1. This listing follows a very similar pattern to that given in Listing 7.1, which we assume the reader is familiar with; the discussion below focuses on the novel features of Listing 8.1. The MATLAB function again requires the parameters N_x, N_y that allow the mesh to be specified as input and returns **x**, **y**, *lnods* and **U**. In lines 4–10, we specify the functions that appear in Eqs. (8.1)–(8.3). The function $g_N(x, y)$ is specified using two separate functions—the function `gn1` specifies the Neumann boundary conditions on $y = 1$, and the function `gn2` specifies the Neumann boundary condition on $x = -1$.

The function `GenerateMesh` is called in line 13 of the listing. The body of this function occupies lines 74–115 of the listing. In addition to the arrays generated in Listing 7.1, this function also generates an array `neu_edges`. This array has two columns, and each row of this array stores the nodes at either end of an element edge that lies on $\partial\Omega_N$. This array is generated in lines 109–115 of the listing.

Having specified the mesh, we then initialise A and **b** (lines 16–17), and loop over all elements (lines 21–29) calculating the nonzero local contributions from integrating over each element, and adding these contributions into the global matrix A and global vector **b**. The local contributions are calculated using the function `CalculateContributionOnElement` that occupies lines 121–190 of the listing. This function first defines the locations of the nodes of the element (lines 126–131), calculates the determinant of the mapping from element k to the canonical element (line 135), and calculates the partial derivatives of basis functions on each element (lines 139–141). We then define the quadrature points and weights in lines 144–146, before initialising the local contributions to A and **b**. We then perform the quadrature required to evaluate the integrals given by Eqs. (8.18) and (8.19), using the quadrature rule given by Eq. (8.27). We loop over the quadrature points, first evaluating the basis functions (lines 157–160), before evaluating the values of x and y at the quadrature points (lines 163–164). We then use these values of x and y to evaluate the required function values (lines 167–171). Finally, in lines 174–190, we use the quadrature rule to calculate the local entries of A and **b**.

Returning to the main body of the code, we loop over all element edges on $\partial\Omega_N$ in lines 34–49, calculating local contributions to **b** using the function `CalculateContributionOnNeumannEdge`, and adding these contributions into the global vector **b**. The function `CalculateContributionOnNeumannEdge` is written in lines 197–233 of the listing. We begin by defining the location of the nodes of the line segment in lines 202–205, before calculating the length of the element in line 208. The quadrature rule is defined in lines 211–213. The local contribution to **b** is initialised in line 216, before being calculated, using the quadrature rule, in lines 229–233.

The Dirichlet boundary conditions are handled by lines 52–56, completing the assembly of the linear system. The linear system is then solved (lines 59–62), and the finite element solution plotted in lines 65–68.

The MATLAB function in Listing 8.1 may be executed by typing, for example,

```
[x, y, lnods, U] = Chap8_CalculateExampleFem(50, 50);
```

into a MATLAB session.

Listing 8.1 `Chap8_CalculateExampleFem.m`, a MATLAB function for calculating the finite element solution of the model problem given by Eqs. (8.1)–(8.3).

```
1  function [x, y, lnods, U] = Chap8_CalculateExampleFem(Nx,Ny)
2  
3  % Define functions
4  p = @(x,y) exp(x);
5  q1 = @(x,y) 3*x;
6  q2 = @(x,y) 2*y;
7  r = @(x,y) -6;
8  f = @(x,y) -2*exp(x)*(1+x+6*y);
9  gn1 = @(x,y) 6*exp(x);
10 gn2 = @(x,y) 2*exp(x);
11 
12 % Generate data structures needed to specify the mesh
13 [x, y, lnods, dir_nodes, neu_edges] = GenerateMesh(Nx, Ny);
14 
15 % Initialise A and b to zero
16 A = sparse((Nx+1)*(Ny+1), (Nx+1)*(Ny+1));
17 b = zeros((Nx+1)*(Ny+1), 1);
18 
19 % Loop over elements, calculating local contributions, and
20 % incrementing global A and b
21 for k=1:2*Nx*Ny
22     [A_local, b_local] = ...
23         CalculateContributionOnElement(x, y, lnods, k, ...
24         p, q1, q2, r, f);
25 
26     A(lnods(k,:), lnods(k,:)) = ...
27         A(lnods(k,:), lnods(k,:)) + A_local;
28     b(lnods(k,:)) = b(lnods(k,:)) + b_local;
29 end
30 
31 % Loop over edges on y=1, calculating local contributions
32 % from Neumann boundary conditions, and adding into the
33 % global b
34 for k=1:Nx
35     b_local = CalculateContributionOnNeumannEdge(x, y, ...
36         neu_edges, k, gn1);
37 
38     b(neu_edges(k,:)) = b(neu_edges(k,:)) + b_local;
39 end
40 
41 % Loop over edges on x=-1, calculating local contributions
42 % from Neumann boundary conditions, and adding into the
43 % global b
44 for k=Nx+1:Nx+Ny
45     b_local = CalculateContributionOnNeumannEdge(x, y, ...
46         neu_edges, k, gn2);
47 
48     b(neu_edges(k,:)) = b(neu_edges(k,:)) + b_local;
49 end
```

```

% Handle Dirichlet boundary conditions
A(dir_nodes,:) = 0;
for i=1:length(dir_nodes)
    A(dir_nodes(i),dir_nodes(i)) = 1;
end
b(dir_nodes) = x(dir_nodes).^2 + 2*y(dir_nodes).^3;

% Compute ILU factorisation of A
[Lower, Upper] = ilu(A);

% Use ILU preconditioned GMRES to solve linear system
U = gmres(A, b, 100, [], 1000, Lower, Upper);

% Plot solution
trisurf(lnods, x, y, U);
xlabel('x')
ylabel('y')
zlabel('U')




% Function to generate data structures required by the mesh
function [x, y, lnods, dir_nodes, neu_edges] = ...
    GenerateMesh(Nx, Ny)

% Set entries of lnods
lnods = zeros(2*Nx*Ny, 3);
for j=1:Ny;
    for i=1:Nx;
        ele = 2*(j-1)*Nx+2*i-1;
        lnods(ele,:) = ...
            [(j-1)*(Nx+1)+i, (j-1)*(Nx+1)+i+1, j*(Nx+1)+i];

        ele = ele+1;
        lnods(ele,:) = ...
            [j*(Nx+1)+i+1, j*(Nx+1)+i, (j-1)*(Nx+1)+i+1];
    end
end

% Set entries of x and y over the interval from 0 to 1
x = zeros((Nx+1)*(Ny+1),1);
y = zeros((Nx+1)*(Ny+1),1);

for i=1:Nx+1
    x(i:Nx+1:(Nx+1)*(Ny+1)) = 2*(i-1)/Nx-1;
end

for j=1:Ny+1
    y((j-1)*(Nx+1)+1:j*(Nx+1)) = 2*(j-1)/Ny-1;
end

% Specify locations of Dirichlet boundary conditions
% Nodes on y=-1
dir_nodes(1:Nx+1) = 1:Nx+1;
% Nodes on x=1
dir_nodes(Nx+2:Nx+Ny+1) = 2*(Nx+1):Nx+1:(Ny+1)*(Nx+1);

% Specify locations of Neumann boundary conditions
% Edges on y=1
neu_edges(1:Nx,:) = [((Ny+1)*(Nx+1):-1:Ny*(Nx+1)+2)', ...
    ((Ny+1)*(Nx+1)-1:-1:Ny*(Nx+1)+1)'];
```

```
% Edges on x=-1
neu_edges(Nx+1:2*Nx,:) = [(Ny*(Nx+1)+1:-(Nx+1):Nx+2)', ...
    ((Ny-1)*(Nx+1)+1:-(Nx+1):1)'];




% Calculate local contributions to A and b from element k
function [A_local, b_local] = ...
    CalculateContributionOnElement(x, y, lnods, k, p, ...
        q1, q2, r, f)

% Define x and y coordinate of each node
x1 = x(lnods(k,1));
x2 = x(lnods(k,2));
x3 = x(lnods(k,3));
y1 = y(lnods(k,1));
y2 = y(lnods(k,2));
y3 = y(lnods(k,3));

% Calculate Jacobian of transformation between canonical
% element and element k
detF = (x2-x1)*(y3-y1) - (x3-x1)*(y2-y1);

% Evaluate derivatives of basis functions in terms of (x,y)
% coordinates
Finv = [y3-y1, x1-x3; y1-y2, x2-x1]/detF;
dphi_dX = [-1, -1; 1, 0; 0, 1];
dphi_dx = dphi_dX*Finv;

% Define quadrature points
M = 3;
QuadWeights = [5/18, 4/9, 5/18];
QuadPoints = [0.5*(1-sqrt(0.6)), 0.5, 0.5*(1+sqrt(0.6))];

% Initialise local A and b to zero
A_local = zeros(3,3);
b_local = zeros(3,1);

% Loop over quadrature points to calculate local ...
% contributions to A and b
for n=1:M
    for m=1:M
        % Set values of basis functions
        phi(1) = 1-QuadPoints(m)- ...
            (1-QuadPoints(m))*QuadPoints(n);
        phi(2) = QuadPoints(m);
        phi(3) = (1-QuadPoints(m))*QuadPoints(n);

        % Set values of x and y
        x_val = x1*phi(1) + x2*phi(2) + x3*phi(3);
        y_val = y1*phi(1) + y2*phi(2) + y3*phi(3);

        % Evaluate functions of x and y
        p_val = p(x_val, y_val);
        q1_val = q1(x_val, y_val);
        q2_val = q2(x_val, y_val);
        r_val = r(x_val, y_val);
        f_val = f(x_val, y_val);

        % Add weighted quadrature value to local A and b
        for i=1:3
            for j=1:3
```

```
                        A_local(i,j) = A_local(i,j) + ...
                            detF*QuadWeights(m)* ...
                            QuadWeights(n)*(1-QuadPoints(m))* ...
                            (p_val*(dphi_dx(i,1)*dphi_dx(j,1) + ...
                            dphi_dx(i,2)*dphi_dx(j,2)) + ...
                            q1_val*phi(i)*dphi_dx(j,1) + ...
                            q2_val*phi(i)*dphi_dx(j,2) + ...
                            r_val*phi(i)*phi(j));
                    end
                    b_local(i) = b_local(i) + ...
                        detF*QuadWeights(m)*QuadWeights(n)* ...
                        (1-QuadPoints(m))*f_val*phi(i);
                end
            end
        end






function b_local = ...
    CalculateContributionOnNeumannEdge(x, y, neu_edges, ...
    k, gn)

% Define x and y coordinate of each node
x1 = x(neu_edges(k,1));
x2 = x(neu_edges(k,2));
y1 = y(neu_edges(k,1));
y2 = y(neu_edges(k,2));

% Calculate length of edge
h = sqrt((x1-x2)^2 + (y1-y2)^2);

% Define quadrature rule
M = 3;
QuadWeights = [5/18, 4/9, 5/18];
QuadPoints = [0.5*(1-sqrt(0.6)), 0.5, 0.5*(1+sqrt(0.6))];

% Initialise local b to zero
b_local = zeros(2,1);

% Loop over quadrature points to calculate local
% contributions to b
for m=1:M
    % Define basis functions at each quadrature point
    phi = [1-QuadPoints(m); QuadPoints(m)];

    % Define values of x and y
    x_val = x1*phi(1) + x2*phi(2);
    y_val = y1*phi(1) + y2*phi(2);

    % Evaluate g_N
    gn_val = gn(x_val, y_val);

    % Add weighted contribution to b
    b_local = b_local + h*QuadWeights(m)*gn_val*phi;
end
```

8.9 Exercises

8.1 Calculate the finite element solution of the boundary value problem, defined for $0 < x < 2, 0 < y < 1$, by

$$-4\left(\frac{\partial^2 u}{\partial x^2} + \frac{\partial^2 u}{\partial y^2}\right) + 4\frac{\partial u}{\partial x} + 2\frac{\partial u}{\partial y} + 2u = 1,$$

subject to the Dirichlet boundary conditions

$$\begin{aligned} u(x,0) &= e^x + \frac{1}{2}, & 0 < x < 2, \\ u(2,y) &= e^{y+2} + \frac{1}{2}, & 0 < y < 1, \\ u(x,1) &= e^{x+1} + \frac{1}{2}, & 0 < x < 2, \\ u(0,y) &= e^y + \frac{1}{2}, & 0 < y < 1. \end{aligned}$$

The solution of this boundary value problem is $u(x, y) = e^{x+y} + \frac{1}{2}$. Use this analytic solution to verify that your finite element solution is correct.

8.2 Calculate the finite element solution of the boundary value problem, defined for $0 < x < 2, -1 < y < 0$, by

$$-\nabla \cdot \left(e^{-x^2-y}\nabla u\right) + e^{-x^2}\frac{\partial u}{\partial x} + e^{-y}\frac{\partial u}{\partial y} - e^{-y}u = 2xe^y - 2,$$

subject to the Dirichlet boundary conditions

$$\begin{aligned} u(x,0) &= e^{x^2}, & 0 < x < 2, \\ u(2,y) &= e^{y+4}, & -1 < y < 0, \\ u(x,-1) &= e^{x^2-1}, & 0 < x < 2, \\ u(0,y) &= e^y, & -1 < y < 0. \end{aligned}$$

The solution of this boundary value problem is $u(x, y) = e^{x^2+y}$. Use this analytic solution to verify that your finite element solution is correct.

8.3 Modify your solution to Exercise 8.2 to calculate the finite element solution of the boundary value problem, defined for $0 < x < 2, -1 < y < 0$, by

$$-\nabla \cdot \left(e^{-x^2-y}\nabla u\right) + e^{-x^2}\frac{\partial u}{\partial x} + e^{-y}\frac{\partial u}{\partial y} - e^{-y}u = 2xe^y - 2,$$

subject to the Dirichlet boundary conditions

$$u(x,0) = \mathrm{e}^{x^2}, \qquad 0 < x < 2,$$
$$u(2,y) = \mathrm{e}^{y+4}, \qquad -1 < y < 0,$$

and the Neumann boundary conditions

$$\mathrm{e}^{-x^2-y}\,(\nabla u)\cdot\mathbf{n} = -1 \qquad 0 < x < 2, \qquad y = -1,$$
$$\mathrm{e}^{-x^2-y}\,(\nabla u)\cdot\mathbf{n} = 0 \qquad x = 0, \qquad -1 < y < 0.$$

Use the true solution, $u(x,y) = \mathrm{e}^{x^2+y}$, to verify that your finite element solution is correct.

8.4 Define the region Ω by the intersection of the regions given by

$$x > 0, \qquad y > 0, \qquad y < 1, \qquad x + y < 11.$$

The partial differential equation is defined by

$$-\left(\frac{\partial^2 u}{\partial x^2} + \frac{\partial^2 u}{\partial y^2}\right) = -4, \qquad (x,y) \in \Omega,$$

subject to the Dirichlet boundary conditions

$$u(x,0) = x^2, \qquad 0 < x < 11,$$
$$u(0,y) = y^2, \qquad 0 < y < 1,$$

and Neumann boundary conditions

$$(\nabla u)\cdot\mathbf{n} = 2, \qquad 0 < x < 10, \qquad y = 1,$$
$$(\nabla u)\cdot\mathbf{n} = \sqrt{2}(x+y), \qquad 10 < x < 11, \qquad x + y = 11.$$

This boundary value problem has solution $u(x,y) = x^2 + y^2$.

(a) Modify the mesh generated in Listing 8.1 so that it covers the region given by $0 < x < 10$, $0 < y < 1$.
(b) Scale the values of x_i at each node in the mesh generated in part (a) by a factor of $1 + 0.1(1 - y_i)$, and verify that this new mesh partitions the region Ω.
(c) Use the mesh generated in part (b) to calculate the finite element solution of the boundary value problem above. Verify that your finite element solution is correct by comparison with the true solution.

8.5 A more general partial differential equation than that given by Eq. (8.1) is defined on some region Ω by

$$-\nabla \cdot (P(x, y)\nabla u) + \mathbf{q}(x, y) \cdot \nabla u + r(x, y)u = f(x, y), \qquad (x, y) \in \Omega,$$

where $P(x, y)$ is a 2×2 matrix with entries given by

$$P(x, y) = \begin{pmatrix} P_{11}(x, y) & P_{12}(x, y) \\ P_{21}(x, y) & P_{22}(x, y) \end{pmatrix},$$

and $P_{11}, P_{12}, P_{21}, P_{22}, \mathbf{q}, r, f$ are given functions. On the boundary $\partial\Omega_D$, we apply Dirichlet boundary conditions:

$$u(x, y) = g_D(x, y), \qquad (x, y) \in \partial\Omega_D,$$

and on the boundary $\partial\Omega_N$, we apply natural Neumann boundary conditions:

$$(P\nabla u) \cdot \mathbf{n} = g_N(x, y), \qquad (x, y) \in \partial\Omega_N,$$

where $\mathbf{n}$ is the normal vector to the boundary $\partial\Omega_N$, pointing out of the domain Ω, and of unit length.

(a) Write down the weak formulation of the boundary value problem above.
(b) Note that, if

$$P(x, y) = \begin{pmatrix} 4 & 0 \\ 0 & 1 \end{pmatrix},$$

then

$$\nabla \cdot (P(x, y)\nabla u) = 4\frac{\partial^2 u}{\partial x^2} + \frac{\partial^2 u}{\partial y^2}.$$

Calculate the finite element solution of the partial differential equation, defined for $0 < x < 1, 0 < y < 1$, by

$$-\left(4\frac{\partial^2 u}{\partial x^2} + \frac{\partial^2 u}{\partial y^2}\right) + \left(2x + 4x^2y^2 + 4y^4\right) u = 0,$$

subject to Dirichlet boundary conditions

$$\begin{aligned} u(x, 0) &= 1, && 0 < x < 1, \\ u(1, y) &= e^{y^2}, && 0 < y < 1, \\ u(x, 1) &= e^x, && 0 < x < 1, \\ u(0, y) &= 1, && 0 < y < 1. \end{aligned}$$

The solution of this differential equation is $u(x, y) = e^{xy^2}$. Use this analytic solution to verify that your finite element solution is correct.

Chapter 9
Quadrilateral Elements

In Chaps. 7 and 8, we calculated the finite element solution of partial differential equations, having partitioned the domain on which the differential equation is defined into a mesh of triangular elements. Triangular elements are not, however, the only shape of elements that can be used when partitioning the domain. In this chapter, we explain how a mesh comprising quadrilateral elements may be used when calculating the finite element solution of a partial differential equation.

9.1 A Model Problem

We use the same model differential equation as in Chap. 8:

$$-\nabla \cdot (p(x, y)\nabla u) + \mathbf{q}(x, y) \cdot \nabla u + r(x, y)u = f(x, y), \qquad (x, y) \in \Omega, \tag{9.1}$$

where Ω is some finite region of the (x, y)-plane, and $p(x, y)$, $\mathbf{q}(x, y)$, $r(x, y)$ and $f(x, y)$ are given functions. The boundary, $\partial\Omega$, is partitioned into two regions, $\partial\Omega_D$ and $\partial\Omega_N$. On the boundary $\partial\Omega_D$, we apply Dirichlet boundary conditions:

$$u(x, y) = g_D(x, y), \qquad (x, y) \in \partial\Omega_D, \tag{9.2}$$

and on the boundary $\partial\Omega_N$, we apply natural Neumann boundary conditions:

$$p(x, y)\,(\nabla u) \cdot \mathbf{n} = g_N(x, y), \qquad (x, y) \in \partial\Omega_N, \tag{9.3}$$

where $\mathbf{n}$ is the normal vector to the boundary $\partial\Omega_N$, pointing out of the domain Ω, and of unit length.

J. Whiteley, *Finite Element Methods*, Mathematical Engineering,
DOI 10.1007/978-3-319-49971-0_9

9.2 The Weak Formulation

The weak formulation of Eqs. (9.1)–(9.3) is identical to that derived in Sect. 8.2 and is given by:

find $u(x, y) \in H_E^1(\Omega)$, such that

$$\int_{\Omega} p\,(\nabla v) \cdot (\nabla u) + v\mathbf{q} \cdot \nabla u + ruv \,\mathrm{d}A = \int_{\Omega} fv \,\mathrm{d}A + \int_{\partial\Omega_N} vg_N \,\mathrm{d}s, \qquad (9.4)$$

for all $v(x, y) \in H_0^1(\Omega)$.

9.3 The Computational Mesh

In the exercises at the end of this chapter, we will develop computational implementations for calculating the finite element solution of boundary value problems. These boundary value problem will be posed on rectangular domains $a < x < b$, $c < y < d$, where a, b, c, d are constants, and so we now explain how to partition a rectangular domain into a mesh of quadrilateral elements.

A rectangular domain may easily be partitioned into a mesh of uniform rectangles, with N_x rectangles in the x-direction and N_y rectangles in the y-direction, as shown in Fig. 9.1 when $N_x = 2$ and $N_y = 3$. As with the mesh of triangular elements developed in Sect. 7.3.1, we number the nodes, starting with the node at the bottom left-hand corner of the mesh, and number the nodes consecutively moving in the x-direction from this first node. We then move up to the next row in the y-direction and repeat this process, starting from the node at the left-hand end of these nodes. We proceed through the mesh in this fashion until all nodes have been numbered. This numbering convention is illustrated in Fig. 9.1a when $N_x = 2$ and $N_y = 3$. A similar process is used to number the elements—this is illustrated in Fig. 9.1b when $N_x = 2$ and $N_y = 3$.

As when using a mesh of triangular elements, we will require a connectivity array, again called $lnods$, to store the nodes of each element. Row k of this array will store the nodes of element k, listed in an anticlockwise direction. Each quadrilateral element will contain four nodes and so, if there are N_{ele} elements in the mesh, this array will be of size $N_{\text{ele}} \times 4$. For the mesh shown in Fig. 9.1, where $N_x = 2$ and $N_y = 3$, the connectivity array may be written as

$$lnods = \begin{pmatrix} 1 & 2 & 5 & 4 \\ 2 & 3 & 6 & 5 \\ 4 & 5 & 8 & 7 \\ 5 & 6 & 9 & 8 \\ 7 & 8 & 11 & 10 \\ 8 & 9 & 12 & 11 \end{pmatrix}.$$

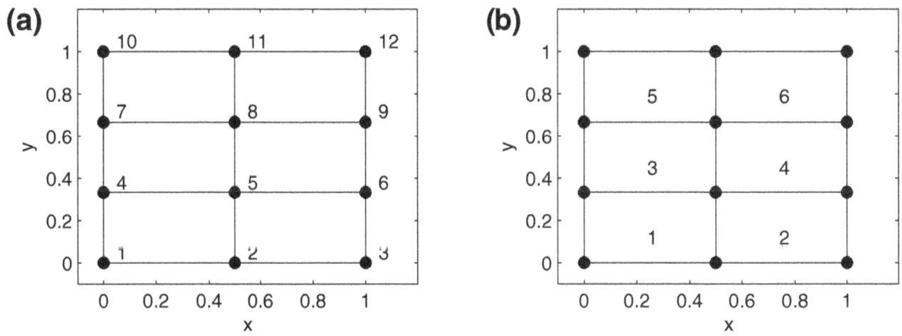

Fig. 9.1 **a** The labelling of the nodes in the mesh. **b** The labelling of the elements in the mesh

In common with a mesh of triangular elements, the connectivity array is not unique. First, the order that the elements of the mesh are stored in does not matter, and so we may interchange any of the rows of this array. Furthermore, we only have to list the nodes of each element in an anticlockwise direction—the first row of the connectivity array above could be written 2, 5, 4, 1, or 5, 4, 1, 2 or 4, 1, 2, 5 instead of 1, 2, 5, 4.

In Sect. 7.3.1, we defined a hanging node to be a point in the domain Ω that is a node of one element, but lies on an edge of another element without being a node. We illustrate an example of a hanging node in a mesh of quadrilateral elements in Fig. 9.2, with the hanging node identified by an open circle. As with triangular elements, we do not permit hanging nodes in a mesh of quadrilateral elements.

Finally, we reiterate the warning about long, thin elements that we first gave in Sect. 7.3.1. Symptoms of long, thin elements include the following: (i) a high ratio between the length of the longest edge and the length of the shortest edge of an element; (ii) a very low interior angle between edges of an element; or (iii) an interior angle between edges of an element that is close to 180°. Elements such as these should be avoided, as they can degrade the accuracy of the finite element solution.

Fig. 9.2 An example of a hanging node

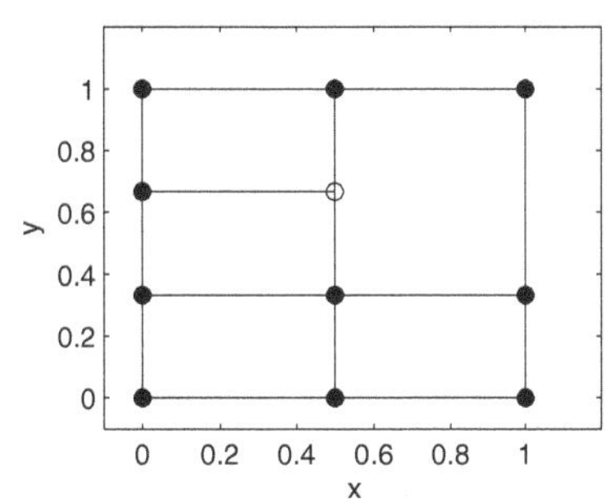

9.4 Basis Functions

In the previous chapters, whether calculating the finite element solution of an ordinary differential equation or a partial differential equation, we defined global basis functions through: (i) a transformation between a general element of the mesh and a canonical element and (ii) the definition of local basis functions on this canonical element. For a mesh of quadrilateral elements, a suitable canonical element is the unit square in (X, Y)-coordinates, defined by $0 \leq X \leq 1, 0 \leq Y \leq 1$, and illustrated in Fig. 9.3.

Suppose element k of the mesh includes nodes the k_1, k_2, k_3, k_4, listed in an anticlockwise direction. A mapping between this element (in (x, y)-coordinates) and the canonical element (in (X, Y)-coordinates) is defined by, for $0 \leq X \leq 1$, $0 \leq Y \leq 1$:

$$x = (1-X)(1-Y)x_{k_1} + X(1-Y)x_{k_2} + XYx_{k_3} + (1-X)Yx_{k_4}, \tag{9.5}$$
$$y = (1-X)(1-Y)y_{k_1} + X(1-Y)y_{k_2} + XYy_{k_3} + (1-X)Yy_{k_4}. \tag{9.6}$$

Local basis functions on the canonical element are defined by, for $0 \leq X \leq 1$, $0 \leq Y \leq 1$,

$$\phi_{\text{local},1}(X, Y) = (1-X)(1-Y), \tag{9.7}$$
$$\phi_{\text{local},2}(X, Y) = X(1-Y), \tag{9.8}$$
$$\phi_{\text{local},3}(X, Y) = XY, \tag{9.9}$$
$$\phi_{\text{local},4}(X, Y) = (1-X)Y. \tag{9.10}$$

The local basis functions defined by Eqs. (9.7)–(9.10) are known as *bilinear* local basis functions, as they are products of a linear function of X and a linear function of Y. They are not, however, linear functions when viewed as a function of both X

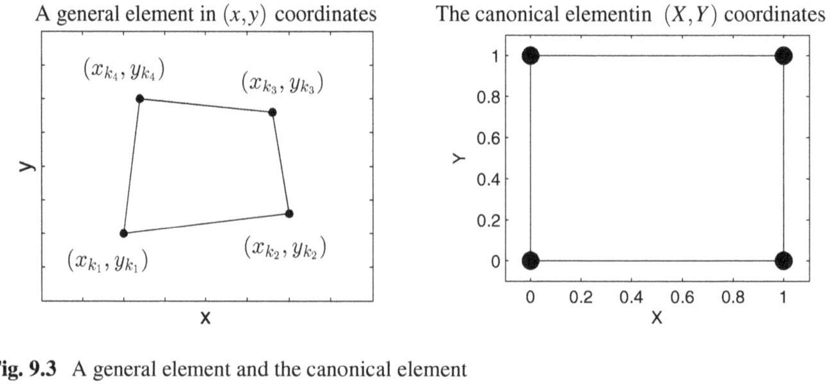

Fig. 9.3 A general element and the canonical element

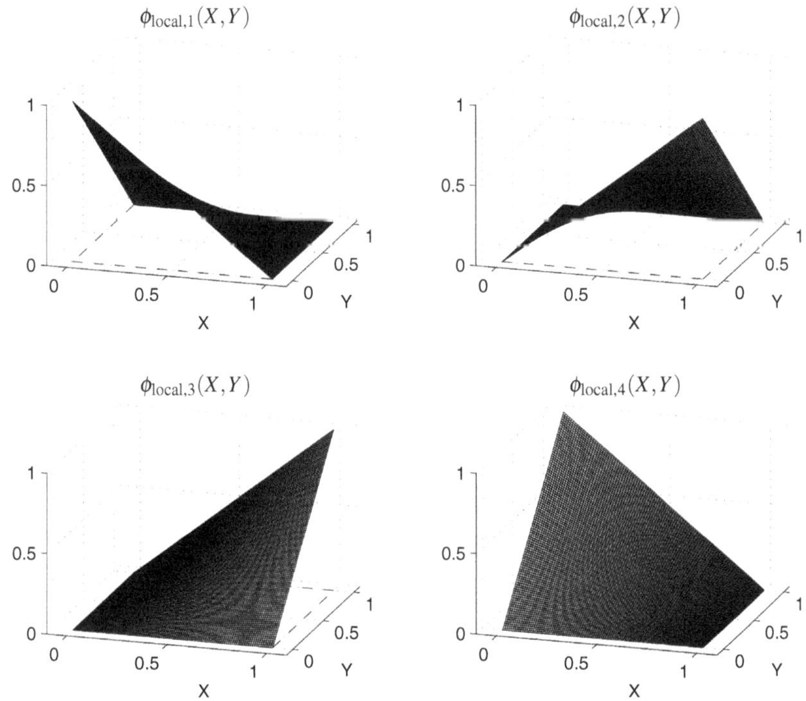

Fig. 9.4 The local basis functions $\phi_{\text{local},1}$, $\phi_{\text{local},2}$, $\phi_{\text{local},3}$, $\phi_{\text{local},4}$ defined by Eqs. (9.7)–(9.10)

and Y however, as they include the quadratic term XY. These local basis functions are illustrated in Fig. 9.4.

Global basis functions are now defined on element k, in terms of the local basis functions given by Eqs. (9.7)–(9.10), by

$$\phi_{k_1}(X, Y) = \phi_{\text{local},1}(X, Y), \tag{9.11}$$

$$\phi_{k_2}(X, Y) = \phi_{\text{local},2}(X, Y), \tag{9.12}$$

$$\phi_{k_3}(X, Y) = \phi_{\text{local},3}(X, Y), \tag{9.13}$$

$$\phi_{k_4}(X, Y) = \phi_{\text{local},4}(X, Y), \tag{9.14}$$

$$\phi_j(X, Y) = 0, \qquad j \neq k_1, j \neq k_2, j \neq k_3, j \neq k_4. \tag{9.15}$$

We illustrate an example global basis function as a function of x and y in Fig. 9.5.

Suppose the mesh contains N_{node} nodes. We leave it as an exercise for the reader to verify whether the global basis functions defined by Eqs. (9.5)–(9.15) satisfy the usual property that, for $j = 1, 2, \ldots, N_{\text{node}}$,

$$\phi_j(x_i) = \begin{cases} 1, & i = j, \\ 0, & i \neq j. \end{cases} \tag{9.16}$$

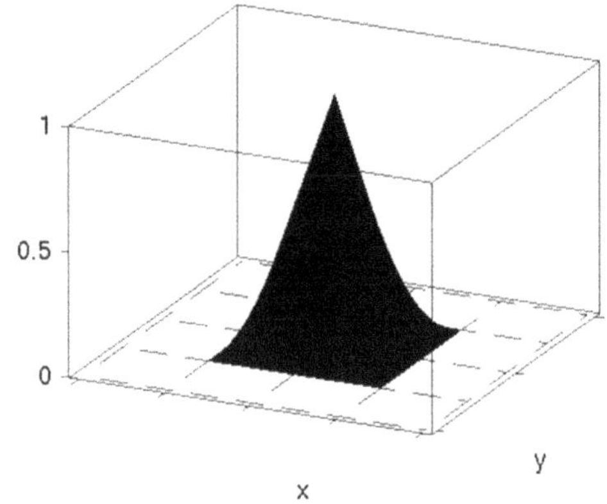

Fig. 9.5 An example global basis function

9.5 Sets of Functions

We now define the sets of functions S^h, S^h_E, S^h_0 that will be used when specifying the finite element solution. We define

$$S^h = \Big\{ \text{functions } V(x, y) \text{ given by } V(x, y) = \sum_{j=1}^{N_{\text{node}}} V_j \phi_j(x, y) \text{ where } \phi_j(x, y), \\ j = 1, 2, \ldots, N_{\text{node}}, \text{ are defined by Eqs. (9.5)–(9.15)} \Big\}.$$

The sets S^h_E and S^h_0 are then defined by

$$S^h_E = \Big\{ \text{functions } V(x, y) \text{ in } S^h \text{ such that } V(x, y) \text{ satisfies all Dirichlet boundary conditions} \Big\},$$

$$S^h_0 = \Big\{ \text{functions } V(x, y) \text{ in } S^h \text{ such that } V(x, y) = 0 \text{ at all points} (x, y) \text{ where Dirichlet boundary conditions are specified} \Big\}.$$

9.6 The Finite Element Formulation

Having defined the weak solution by Eq. (9.4), the sets S^h_E and S^h_0 defined in Sect. 9.5 allow us to define the finite element solution by:

find $U(x, y) \in S_E^h$, such that

$$\int_\Omega p\,(\nabla\phi_i)\cdot(\nabla U) + \phi_i \mathbf{q}\cdot\nabla U + rU\phi_i \,\mathrm{d}A = \int_\Omega f\phi_i \,\mathrm{d}A + \int_{\partial\Omega_N} \phi_i g_N \,\mathrm{d}s, \tag{9.17}$$

for all $\phi_i(x, y) \subset S_0^h$.

9.7 The System of Algebraic Equations

The finite element solution is written in the usual form:

$$U(x, y) = \sum_{j=1}^{N_{\text{node}}} U_j\phi_j(x, y), \tag{9.18}$$

where N_{node} is the number of nodes in the mesh. It is easy to see that $U(x, y)$, as defined above, lies in the set S^h.

Although the basis functions, $\phi_1, \phi_2, \ldots, \phi_{N_{\text{node}}}$, used in this chapter are different from the basis functions used in Chap. 8, both the form of the finite element solution, Eq. (9.18), and the finite element formulation, Eq. (9.17), are identical to those given in Eqs. (8.7) and (8.8) in Chap. 8. As a consequence, we may use the same arguments as in Sect. 8.5.3 to deduce that the algebraic equations satisfied by $U_1, U_2, \ldots, U_{N_{\text{node}}}$ may be written as:

$$A\mathbf{U} = \mathbf{b}, \tag{9.19}$$

where, if node i lies on $\partial\Omega_D$,

$$A_{i,j} = \begin{cases} 1, & j = i, \\ 0, & j \neq i, \end{cases} \tag{9.20}$$

$$b_i = g_D(x_i, y_i), \tag{9.21}$$

and if node i does not lie on $\partial\Omega_D$,

$$A_{i,j} = \int_\Omega p\,(\nabla\phi_i)\cdot\left(\nabla\phi_j\right) + \phi_i\mathbf{q}\cdot\left(\nabla\phi_j\right) + r\phi_i\phi_j \,\mathrm{d}A, \qquad j = 1, 2, \ldots, N_{\text{node}}, \tag{9.22}$$

$$b_i = \int_\Omega f\phi_i \,\mathrm{d}A + \int_{\partial\Omega_N} \phi_i g_N \,\mathrm{d}s. \tag{9.23}$$

The entries of A and $\mathbf{b}$ given by Eqs. (9.22) and (9.23) are therefore written in an identical form to those given by Eqs. (8.14) and (8.15). They differ, however, as the basis functions used in this chapter are different to those used in Chap. 8.

9.8 Assembling the System of Algebraic Equations

We will follow the usual procedure when calculating the entries of the linear system given by Eq. (9.19)—that is, we first evaluate the entries, given by Eqs. (9.22) and (9.23), that are given by integrals over the domain, Ω, and integrals over the part of the boundary where Neumann boundary conditions are applied, $\partial\Omega_N$. As before, the integrals over Ω will be calculated by decomposing the integral into the sum of integrals over individual elements, and the integrals over $\partial\Omega_N$ will be calculated by decomposing the integral into the sum of integrals over element edges that lie on $\partial\Omega_N$. We therefore write Eqs. (9.22) and (9.23), the equations for row i of the linear system given by Eq. (9.19), where node i of the mesh does not lie on $\partial\Omega_D$, as

$$A_{i,j} = \sum_{k=1}^{N_{\text{ele}}} \int_{e_k} p\,(\nabla\phi_i)\cdot\left(\nabla\phi_j\right) + \phi_i\mathbf{q}\cdot\left(\nabla\phi_j\right) + r\phi_i\phi_j \;\mathrm{d}A, \qquad j = 1, 2, \ldots, N_{\text{node}}, \tag{9.24}$$

$$b_i = \sum_{k=1}^{N_{\text{ele}}} \int_{e_k} f\phi_i \;\mathrm{d}A + \sum_{k=1}^{N_{\text{Neu}}} \int_{\partial e_{N_k}} \phi_i g_N \;\mathrm{d}s, \tag{9.25}$$

where N_{ele} is the number of elements in the mesh, e_k is the region occupied by element k, N_{Neu} is the number of element edges that lie on $\partial\Omega_N$, and ∂e_{N_k} is edge k of the element edges that lie on $\partial\Omega_N$. We note that Eqs. (9.24) and (9.25) appear identical to Eqs. (8.16) and (8.17). The difference is that the basis functions $\phi_i(x, y)$ are now defined by Eqs. (9.5)–(9.15) rather than Eqs. (7.12)–(7.20). We may, however, assemble these entries of A and $\mathbf{b}$ in the same fashion as in Sects. 8.6.1 and 8.6.2—that is, we first calculate the local contributions from integrating over each element, before calculating the local contributions from integrating along each element edge that lies on $\partial\Omega_N$.

Having assembled the entries of the linear system that are of the form given by Eqs. (9.22) and (9.23), we then set the entries given by Eqs. (9.20) and (9.21) to complete the assembly of the linear system. We now describe each of these steps in a little more detail.

9.8.1 Evaluating Integrals Over Ω

We begin by calculating the contributions to Eqs. (9.24) and (9.25) that arise from integrating over each element. Suppose element k contains the nodes k_1, k_2, k_3, k_4, listed in an anticlockwise order. The only basis functions that are nonzero on this element are $\phi_{k_1}(x)$, $\phi_{k_2}(x)$, $\phi_{k_3}(x)$, $\phi_{k_4}(x)$, and so the nonzero contributions to Eq. (9.24) from integrating over this element can be stored in a matrix $A^{(k)}_{\text{local}}$ of size 4×4, with entries given by, for $i = 1, 2, 3, 4$, $j = 1, 2, 3, 4$:

$$A^{(k)}_{\text{local},i,j} = \int_{e_k} p(x,y)\left(\nabla\phi_{k_i}\right)\cdot\left(\nabla\phi_{k_j}\right) + \phi_{k_i}\mathbf{q}(x,y)\cdot\left(\nabla\phi_{k_j}\right) + r(x,y)\phi_{k_i}\phi_{k_j}\,\mathrm{d}A,$$

where entry $A^{(k)}_{\text{local},i,j}$ contributes to entry A_{k_i,k_j} of the global matrix A.

Similarly, the only nonzero contributions to Eq. (9.25) from integrating over this element may be stored in a vector with four entries, $\mathbf{b}^{(k)}_{\text{local}}$, with entries given by, for $i = 1, 2, 3, 4$:

$$b^{(k)}_{\text{local},i} = \int_{e_k} f(x,y)\phi_{k_i}\,\mathrm{d}A,$$

where entry $b^{(k)}_{\text{local},i}$ contributes to entry b_{k_i} of the global vector $\mathbf{b}$.

As in earlier chapters, we will find it convenient to write these local contributions as integrals over the canonical element, in this case the unit square given by $0 \le X \le 1$, $0 \le Y \le 1$. Using the transformation between element k and the canonical element given by Eqs. (9.5)–(9.6), and the definition of local basis functions given by Eqs. (9.11)–(9.14), we may write, for $i = 1, 2, 3, 4$, $j = 1, 2, 3, 4$:

$$\begin{aligned} A^{(k)}_{\text{local},i,j} = \int_{Y=0}^{1}\Bigg(\int_{X=0}^{1}\Bigg(p(x,y)\left(\frac{\partial\phi_{\text{local},i}}{\partial x}\frac{\partial\phi_{\text{local},j}}{\partial x} + \frac{\partial\phi_{\text{local},i}}{\partial y}\frac{\partial\phi_{\text{local},j}}{\partial y}\right) + \\ \phi_{\text{local},i}\left(q_1(x,y)\frac{\partial\phi_{\text{local},j}}{\partial x} + q_2(x,y)\frac{\partial\phi_{\text{local},j}}{\partial y}\right) + \\ r(x,y)\phi_{\text{local},i}\phi_{\text{local},j}\Bigg)\det(F)\,\mathrm{d}X\Bigg)\,\mathrm{d}Y, \end{aligned} \tag{9.26}$$

where F is the Jacobian matrix of the transformation, defined by

$$F = \begin{pmatrix} \frac{\partial x}{\partial X} & \frac{\partial x}{\partial Y} \\ \frac{\partial y}{\partial X} & \frac{\partial y}{\partial Y} \end{pmatrix},$$

which may be computed using the definitions of x and y given by Eqs. (9.5)–(9.6).

To evaluate the integral that appears in Eq. (9.26), we must first write all terms in the integrand as functions of X and Y. We may use the chain rule to evaluate the partial derivatives of the basis functions that appear in the integral, as was done for basis functions on a triangular element in Eq. (7.39):

$$\begin{aligned}
\begin{pmatrix}
\frac{\partial \phi_{\text{local},1}}{\partial x} & \frac{\partial \phi_{\text{local},1}}{\partial y} \\
\frac{\partial \phi_{\text{local},2}}{\partial x} & \frac{\partial \phi_{\text{local},2}}{\partial y} \\
\frac{\partial \phi_{\text{local},3}}{\partial x} & \frac{\partial \phi_{\text{local},3}}{\partial y} \\
\frac{\partial \phi_{\text{local},4}}{\partial x} & \frac{\partial \phi_{\text{local},4}}{\partial y}
\end{pmatrix}
&=
\begin{pmatrix}
\frac{\partial \phi_{\text{local},1}}{\partial X} & \frac{\partial \phi_{\text{local},1}}{\partial Y} \\
\frac{\partial \phi_{\text{local},2}}{\partial X} & \frac{\partial \phi_{\text{local},2}}{\partial Y} \\
\frac{\partial \phi_{\text{local},3}}{\partial X} & \frac{\partial \phi_{\text{local},3}}{\partial Y} \\
\frac{\partial \phi_{\text{local},4}}{\partial X} & \frac{\partial \phi_{\text{local},4}}{\partial Y}
\end{pmatrix}
\begin{pmatrix}
\frac{\partial X}{\partial x} & \frac{\partial X}{\partial y} \\
\frac{\partial Y}{\partial x} & \frac{\partial Y}{\partial y}
\end{pmatrix} \\
&=
\begin{pmatrix}
-(1-Y) & -(1-X) \\
1-Y & -X \\
Y & X \\
-Y & 1-X
\end{pmatrix} F^{-1},
\end{aligned} \tag{9.27}$$

where we have used the definition of local basis functions given by Eqs. (9.7)–(9.10) to evaluate the partial derivatives of these local basis functions with respect to X and Y, and noted that

$$\begin{aligned}
\begin{pmatrix}
\frac{\partial X}{\partial x} & \frac{\partial X}{\partial y} \\
\frac{\partial Y}{\partial x} & \frac{\partial Y}{\partial y}
\end{pmatrix}
&=
\begin{pmatrix}
\frac{\partial x}{\partial X} & \frac{\partial x}{\partial Y} \\
\frac{\partial y}{\partial X} & \frac{\partial y}{\partial Y}
\end{pmatrix}^{-1} \\
&= F^{-1}.
\end{aligned}$$

On using Eq. (9.27), we may now express all the partial derivatives that appear in Eq. (9.26) as functions of X and Y. Note that, in contrast to the partial derivatives of linear basis functions in Chaps. 7 and 8, the partial derivatives of the bilinear basis functions used in this chapter are not constant on each element.

Functions such as $p(x, y)$ may be evaluated as functions of X and Y by first using Eqs. (9.5)–(9.6) to calculate x and y as functions of X and Y, before using these expressions for x and y to evaluate $p(x, y)$. We may now evaluate all terms that appear in the integrand of Eq. (9.26) as functions of X and Y. If this integral cannot be evaluated analytically, we may use quadrature to approximate its value numerically; we explain in Sect. 9.9 how quadrature may be performed over the unit square $0 \le X \le 1, 0 \le Y \le 1$.

Using the same change of variable as above, we may write the local nonzero contributions to $\mathbf{b}$ from element k as, for $i = 1, 2, 3, 4$:

$$b^{(k)}_{\text{local},i} = \int_{Y=0}^{1} \left(\int_{X=0}^{1} f(x, y)\ \phi_{\text{local},i}(X, Y)\ \det(F)\ \mathrm{d}X \right) \mathrm{d}Y. \tag{9.28}$$

The function $f(x, y)$ may be evaluated on the canonical element as described earlier for functions such as $p(x, y)$. The integral above may then be evaluated analytically, or approximated numerically, using quadrature.

9.8.2 Evaluating Integrals Over Ω_N

Having assembled the contributions to Eqs. (9.24) and (9.25) from integration over elements, we now assemble the contribution to (9.25) from integration over element edges that lie on $\partial\Omega_N$. This may be done in an identical fashion to that described in Sect. 8.6.2.

9.8.3 Setting Entries Defined Explicitly

Finally, we set the entries of A and $\mathbf{b}$ given by Eqs. (9.20) and (9.21). At this stage, it is important to correct any incorrect entries that have been introduced into these rows earlier in the assembly process—see Sect. 8.6.3 for more details.

9.9 Quadrature

When calculating the contributions to the global matrix A and global vector $\mathbf{b}$ that are given by Eqs. (9.26) and (9.28), we have to evaluate integrals of the form

$$\int_{Y=0}^{1}\left(\int_{X=0}^{1} h(X,Y)\,\mathrm{d}X\right)\mathrm{d}Y. \tag{9.29}$$

Writing this double integral as

$$\int_{Y=0}^{1}\left(\int_{X=0}^{1} h(X,Y)\,\mathrm{d}X\right)\mathrm{d}Y = \int_{Y=0}^{1} \hat{h}(Y)\,\mathrm{d}Y, \tag{9.30}$$

where

$$\hat{h}(Y) = \int_{X=0}^{1} h(X,Y)\,\mathrm{d}X, \tag{9.31}$$

we will now show that we may evaluate Eq. (9.29) using numerical quadrature, by adapting the material presented in Sect. 3.5. Suppose we have a quadrature rule for integrating over the interval $0 \leq Z \leq 1$ that uses m quadrature points $Z_1, Z_2, \ldots, Z_m$, with weights $w_1, w_2, \ldots, w_M$. We may then write Eq. (9.31) as

$$\hat{h}(Y) = \sum_{m=1}^{M} w_m h(Z_m, Y).$$

Substituting this into Eq. (9.30) gives

$$\int_{Y=0}^{1}\left(\int_{X=0}^{1} h(X,Y)\,\mathrm{d}X\right)\mathrm{d}Y = \sum_{m=1}^{M} w_m \int_{Y=0}^{1} h(Z_m, Y)\,\mathrm{d}Y,$$

where we have interchanged the order of integration and summation. We then apply the same quadrature rule to the integral over $0 \le Y \le 1$ on the right-hand side of the equation above to obtain

$$\int_{Y=0}^{1}\left(\int_{X=0}^{1} h(X,Y)\,\mathrm{d}X\right)\mathrm{d}Y = \sum_{m=1}^{M}\sum_{n=1}^{M} w_m w_n h(Z_m, Z_n),$$

which may be used to approximate the integrals given by Eqs. (9.26) and (9.28).

9.10 Exercises

We suggest that the reader begins with the exercise below. Other suitable exercises are then the exercises at the end of Chap. 8, where a finite element solution to the boundary value problems given in these exercises can be calculated using a mesh of quadrilateral elements and bilinear basis functions.

9.1 In Listing 8.1, we provided a computational implementation for calculating the finite element solution, using linear basis functions on a mesh of triangular elements, for the boundary value problem defined, for $-1 < x < 1$, $-1 < y < 1$, by

$$-\nabla \cdot \left(\mathrm{e}^x \nabla u\right) + 3x\frac{\partial u}{\partial x} + 2y\frac{\partial u}{\partial y} - 6u = -2\mathrm{e}^x\left(1 + x + 6y\right),$$

subject to Dirichlet boundary conditions

$$\begin{aligned} u(x,-1) &= x^2 - 2, && -1 < x < 1, \\ u(1,y) &= 2y^3 + 1, && -1 < y < 1, \end{aligned}$$

and natural Neumann boundary conditions

$$\begin{aligned} \mathrm{e}^x\left(\nabla u\right)\cdot \mathbf{n} &= 6\mathrm{e}^x, && -1 < x < 1, && y = 1, \\ \mathrm{e}^x\left(\nabla u\right)\cdot \mathbf{n} &= 2\mathrm{e}^x, && x = -1, && -1 < y < 1. \end{aligned}$$

This boundary value problem has solution $u(x,y) = x^2 + 2y^3$.

In this exercise, we will modify Listing 8.1 so that it solves the boundary value problem using bilinear global basis functions.

(a) Lines 78–89 of Listing 8.1 generate the connectivity array *lnods* for a mesh of triangular elements. Modify this function so that it generates the connectivity array for a mesh of quadrilateral elements, discussed in Sect. 9.3.

(b) Modify the function `CalculateContributionOnElement` so that it computes the nonzero local contributions to the linear system that arise from integrating over element k, $A_{\text{local}}^{(k)}$ and $\mathbf{b}_{\text{local}}^{(k)}$, discussed in Sect. 9.8.1. When modifying this function, you should remember the following: (i) there are now four nodes in each element, and so there will be four local basis functions, and $A_{\text{local}}^{(k)}$ and $\mathbf{b}_{\text{local}}^{(k)}$ will be of a different size; (ii) expressions for the Jacobian of the transformation F and local basis functions will be different; (iii) the Jacobian and partial derivatives of the basis functions will no longer be constant over the element; and (iv) the quadrature rule for integrating over the canonical square element is different to the quadrature rule for integrating over the canonical triangle that is used in Listing 8.1.

(c) Make any other modifications that are needed, such as the number of elements that are looped over in lines 21–29 of Listing 8.1, and verify your implementation is correct by comparison with the true solution.

Chapter 10
Higher Order Basis Functions

In Chap. 4, we described the use of quadratic, and higher order, basis functions when calculating the finite element solution of ordinary differential equations. In this chapter, we will explain how higher order basis functions may be used to calculate the finite element solution of a partial differential equation. We give a concrete example of the use of quadratic basis functions on a mesh of triangular elements. The exercises at the end of the chapter contain examples of higher order basis functions on meshes of both triangular and quadrilateral elements.

10.1 The Model Boundary Value Problem and Weak Formulation

We use the same model differential equation as in Chaps. 8 and 9:

$$-\nabla \cdot (p(x, y)\nabla u) + \mathbf{q}(x, y) \cdot \nabla u + r(x, y)u = f(x, y), \qquad (x, y) \in \Omega, \quad (10.1)$$

where Ω is some finite region of the (x, y)-plane, and $p(x, y)$, $\mathbf{q}(x, y)$, $r(x, y)$ and $f(x, y)$ are given functions. The boundary, $\partial\Omega$, is partitioned into two regions, $\partial\Omega_D$ and $\partial\Omega_N$. On the boundary $\partial\Omega_D$, we apply Dirichlet boundary conditions:

$$u(x, y) = g_D(x, y), \qquad (x, y) \in \partial\Omega_D, \quad (10.2)$$

and on the boundary $\partial\Omega_N$, we apply natural Neumann boundary conditions:

$$p(x, y)\,(\nabla u) \cdot \mathbf{n} = g_N(x, y), \qquad (x, y) \in \partial\Omega_N, \quad (10.3)$$

J. Whiteley, *Finite Element Methods*, Mathematical Engineering,
DOI 10.1007/978-3-319-49971-0_10

where $\mathbf{n}$ is the normal vector to the boundary $\partial\Omega_N$, pointing out of the domain Ω and of unit length. The weak solution, derived in Sect. 8.2, is defined by:

find $u(x, y) \in H_E^1(\Omega)$, such that

$$\int_\Omega p\,(\nabla v) \cdot (\nabla u) + v\mathbf{q} \cdot \nabla u + ruv \,\mathrm{d}A = \int_\Omega fv \,\mathrm{d}A + \int_{\partial\Omega_N} vg_N \,\mathrm{d}s, \quad (10.4)$$

for all $v(x, y) \in H_0^1(\Omega)$.

10.2 Quadratic Basis Functions on a Mesh of Triangular Elements

We partition the mesh into triangular elements, as described in Sect. 7.3.1. A quadratic approximation to the solution on an element will take the form

$$a_1 + a_2x + a_3y + a_4x^2 + a_5xy + a_6y^2,$$

and will be specified by the six values a_i, $i = 1, 2, \ldots, 6$. As we have six degrees of freedom on each element, we will require six basis functions—and therefore six nodes—on each element. Each triangular element of the mesh used in Sect. 7.3.1 contained three nodes, one node at each vertex. By placing an extra node at the mid-point of each edge of the triangle, as shown in Fig. 10.1, we have then assigned six nodes to each element of the mesh. The global numbering of the nodes is shown in Fig. 10.1; the numbering of the elements is identical to that shown in Fig. 7.2b.

We will require a connectivity array, again denoted by $lnods$, for the elements and nodes in the mesh. When using linear basis functions, row k of this array stored the nodes contained by element k, listed in an anticlockwise direction. For quadratic basis functions, the first three entries of row k of the connectivity array will contain the three nodes at the vertices of element k, again listed in an anticlockwise direction. The next three entries of row k will contain the three nodes at the mid-points of each edge of this element, listed in an anticlockwise order starting from the node opposite the first vertex listed. For example, the row of the connectivity array corresponding to the element shown in Fig. 10.2 could be

$$\begin{aligned} &k_1,\ k_2,\ k_3,\ k_4,\ k_5,\ k_6, \\ \text{or}\ &k_2,\ k_3,\ k_1,\ k_5,\ k_6,\ k_4, \\ \text{or}\ &k_3,\ k_1,\ k_2,\ k_6,\ k_4,\ k_5. \end{aligned}$$

The connectivity array for the mesh shown in Fig. 10.1 could be written as

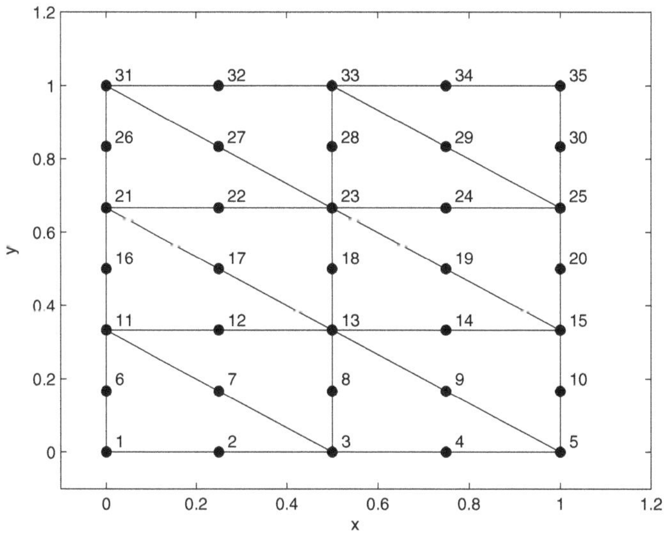

Fig. 10.1 The nodes and elements of a mesh of triangular elements that may be used with quadratic basis functions

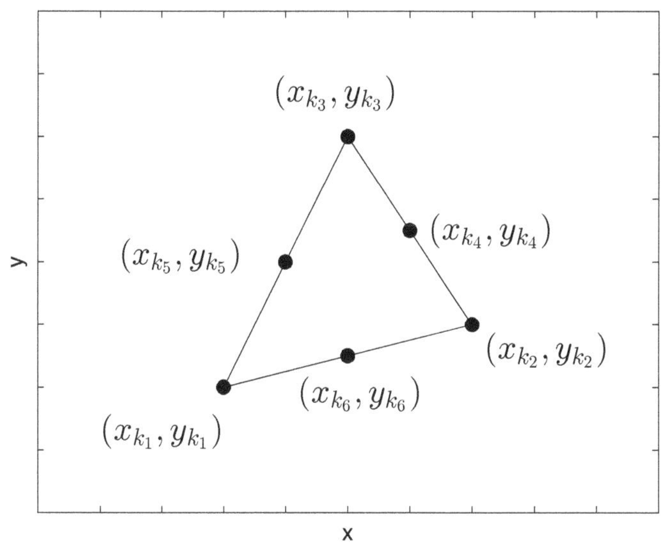

Fig. 10.2 The nodes of an element

$$lnods = \begin{pmatrix} 1 & 3 & 11 & 7 & 6 & 2 \\ 13 & 11 & 3 & 7 & 8 & 12 \\ 3 & 5 & 13 & 9 & 8 & 4 \\ 15 & 13 & 5 & 9 & 10 & 14 \\ 11 & 13 & 21 & 17 & 16 & 12 \\ 23 & 21 & 13 & 17 & 18 & 22 \\ 13 & 15 & 23 & 19 & 18 & 14 \\ 25 & 23 & 15 & 19 & 20 & 24 \\ 21 & 23 & 31 & 27 & 26 & 22 \\ 33 & 31 & 23 & 27 & 28 & 32 \\ 23 & 25 & 33 & 29 & 28 & 24 \\ 35 & 33 & 25 & 29 & 30 & 34 \end{pmatrix}.$$

Global basis functions will, as in Chaps. 7 and 8, be defined by mapping each element of this mesh onto the canonical triangular element and defining local basis functions on this canonical element. Suppose the nodes of element k are $k_1, k_2, k_3, k_4, k_5, k_6$, listed in the order prescribed above for the connectivity array $lnods$. We again map between element k and the canonical element $X \geq 0,\ Y \geq 0,\ 1 - X - Y \geq 0$ using the usual transformation:

$$x = (1 - X - Y)x_{k_1} + Xx_{k_2} + Yx_{k_3}, \tag{10.5}$$
$$y = (1 - X - Y)y_{k_1} + Xy_{k_2} + Yy_{k_3}. \tag{10.6}$$

Local quadratic basis functions $\phi_{\text{local,i}}(X, Y)$, $i = 1, 2, \ldots, 6$ are defined on the canonical element by

$$\phi_{\text{local},1}(X, Y) = 2(1 - X - Y)\left(\frac{1}{2} - X - Y\right), \tag{10.7}$$
$$\phi_{\text{local},2}(X, Y) = 2X\left(X - \frac{1}{2}\right), \tag{10.8}$$
$$\phi_{\text{local},3}(X, Y) = 2Y\left(Y - \frac{1}{2}\right), \tag{10.9}$$
$$\phi_{\text{local},4}(X, Y) = 4XY, \tag{10.10}$$
$$\phi_{\text{local},5}(X, Y) = 4Y(1 - X - Y), \tag{10.11}$$
$$\phi_{\text{local},6}(X, Y) = 4X(1 - X - Y). \tag{10.12}$$

These local basis functions are shown in Fig. 10.3.

Global basis functions are now defined by

$$\phi_{k_i}(X, Y) = \phi_{\text{local,i}}(X, Y), \qquad i = 1, 2, \ldots, 6, \tag{10.13}$$
$$\phi_j(X, Y) = 0, \qquad j \neq k_i, \quad i = 1, 2, \ldots, 6. \tag{10.14}$$

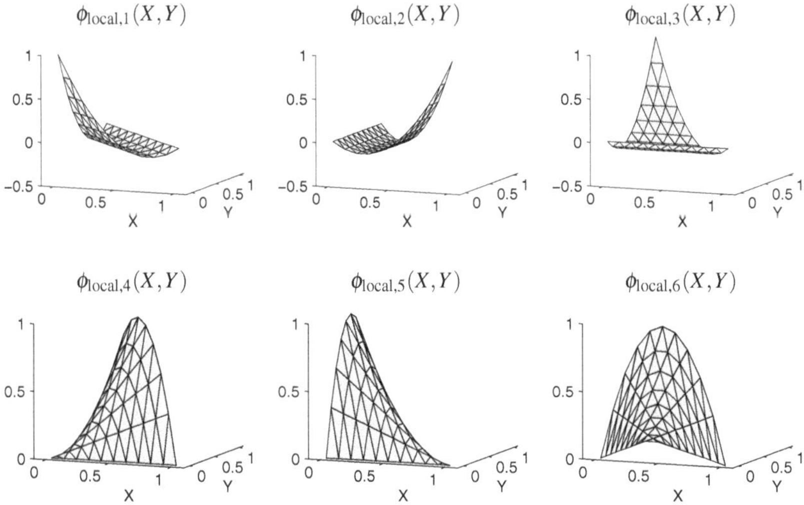

Fig. 10.3 The local basis functions

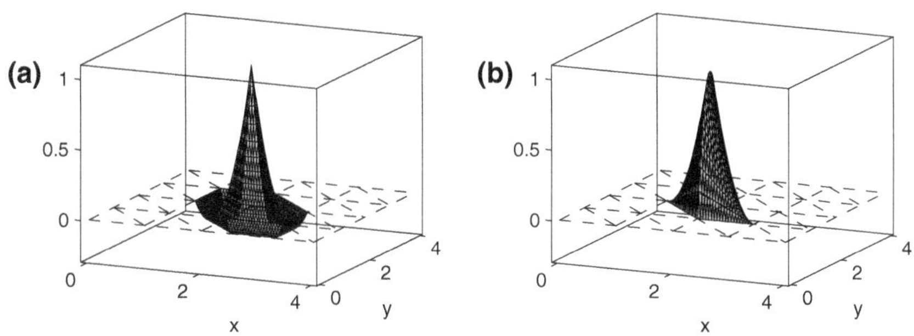

Fig. 10.4 Some example basis functions

Some example basis functions are shown in Fig. 10.4. In Fig. 10.4a, we show an example basis function that takes the value 1 at a node that is at the vertex of a triangle. In Fig. 10.4b, we show an example basis function that takes the value 1 at a node that is at the mid-point of the edge of a triangle.

Remember that nodes that are not at a vertex of a triangular element are placed at the mid-point of an edge of an element. Suppose the mesh contains N_{node} nodes. It is then straightforward to verify that the basis functions given by Eqs. (10.5)–(10.14) satisfy, for $j = 1, 2, \ldots, N_{\text{node}}$, the usual condition:

$$\phi_j(x_i) = \begin{cases} 1, & i = j, \\ 0, & i \neq j. \end{cases} \tag{10.15}$$

10.3 The Finite Element Solution

The sets of functions S^h, S^h_E and S^h_0 are now defined by

$$S^h = \Big\{\text{functions } V(x, y) \text{ given by } V(x, y) = \sum_{j=1}^{N_{\text{node}}} V_j\phi_j(x, y), \text{ where } \phi_j(x, y),$$
$$j = 1, 2, \ldots, N_{\text{node}}, \text{ are defined by Eqs. (10.5)–(10.14)}\Big\},$$
$$S^h_E = \Big\{\text{functions } V(x, y) \text{ in } S^h \text{ such that } V(x, y) \text{ satisfies all Dirichlet boundary conditions}\Big\},$$
$$S^h_0 = \Big\{\text{functions } V(x, y) \text{ in } S^h \text{ such that } V(x, y) = 0 \text{ at all points } (x, y) \text{ where Dirichlet boundary conditions are specified}\Big\}.$$

These sets, and the weak formulation of the problem given by Eq. (10.4), allow us to define the finite element solution by

find $U(x, y) \in S^h_E$, such that

$$\int_\Omega p\,(\nabla\phi_i)\cdot(\nabla U) + \phi_i \mathbf{q}\cdot\nabla U + rU\phi_i \,\mathrm{d}A = \int_\Omega f\phi_i \,\mathrm{d}A + \int_{\partial\Omega_N} \phi_i g_N \,\mathrm{d}s, \tag{10.16}$$

for all $\phi_i(x, y) \in S^h_0$.

10.4 The Algebraic Equations

We write the finite element solution as a member of the set S^h given by

$$U(x, y) = \sum_{j=1}^{N_{\text{node}}} U_j\phi_j(x, y), \tag{10.17}$$

where N_{node} is the number of nodes in the mesh. As explained in Sects. 8.5.3 and 9.7, the algebraic equations satisfied by $U_1, U_2, \ldots, U_{N_{\text{node}}}$ may be written as

$$A\mathbf{U} = \mathbf{b}, \tag{10.18}$$

where, if node i lies on $\partial\Omega_D$, to ensure that $U(x, y) \in S_E^h$, we have

$$A_{i,j} = \begin{cases} 1, & j = i, \\ 0, & j \neq i, \end{cases} \tag{10.19}$$

$$b_i = g_D(x_i, y_i), \tag{10.20}$$

and if node i does not lie on $\partial\Omega_D$,

$$A_{i,j} = \int_\Omega p\,(\nabla\phi_i) \cdot (\nabla\phi_j) + \phi_i \mathbf{q} \cdot (\nabla\phi_j) + r\phi_i\phi_j \;\mathrm{d}A, \qquad j = 1, 2, \ldots, N_{\text{node}}, \tag{10.21}$$

$$b_i = \int_\Omega f\phi_i \;\mathrm{d}A + \int_{\partial\Omega_N} \phi_i g_N \;\mathrm{d}s. \tag{10.22}$$

10.5 Assembling the Algebraic Equations

The entries of the linear system given by Eq. (10.18) are assembled in the same way as they have been throughout this book. We first assemble the entries given by Eqs. (10.21) and (10.22). There are two components to handle when assembling these entries. First, we calculate the terms in Eqs. (10.21) and (10.22) that arise from integrating over Ω. We then calculate the terms in Eq. (10.22) that arise from integrating over $\partial\Omega_N$. Having calculated the entries given by Eqs. (10.21) and (10.22), we then set the entries given by Eqs. (10.19) and (10.20). We now give more details on how these tasks are performed for quadratic basis functions. We assume that the reader is familiar with the assembly of the linear system described in Sect. 8.6 when linear basis functions were used and focus on the description of additional features that arise when using quadratic basis functions.

10.5.1 Evaluating Integrals Over Ω

We calculate the integrals over Ω that appear in Eqs. (10.21) and (10.22) by summing the contributions to these integrals from individual elements. Suppose the nodes of element k are $k_1, k_2, k_3, k_4, k_5, k_6$, listed in the order prescribed for the connectivity array *lnods*. Only the basis functions ϕ_{k_i}, $i = 1, 2, \ldots, 6$, will be nonzero on this element, and so the nonzero local contributions to Eqs. (10.21) and (10.22) from integration over this element may be stored in a 6×6 matrix $A^{(k)}_{\text{local}}$ and a vector $\mathbf{b}^{(k)}_{\text{local}}$, with six entries. Using Eqs. (10.5)–(10.14), we may write the entries of $A^{(k)}_{\text{local},i,j}$ as integrals over the canonical element given by, for $i = 1, 2, \ldots, 6$, $j = 1, 2, \ldots, 6$:

$$A^{(k)}_{\text{local},i,j} = \int_{\triangle} \Big(p(x,y) \left(\nabla \phi_{\text{local},i}\right) \cdot \left(\nabla \phi_{\text{local},j}\right) + \phi_{\text{local},i} \mathbf{q}(x,y) \cdot \left(\nabla \phi_{\text{local},j}\right) + r(x,y) \phi_{\text{local},i} \phi_{\text{local},j} \Big) \det(F) \, \mathrm{d}A_X, \tag{10.23}$$

where, as in Chap. 7, $\int_{\triangle}$, together with $\mathrm{d}A_X$, denotes integration over the canonical triangular element, and $\det(F)$ is the Jacobian of the transformation between element k and the canonical element, given by Eq. (7.37). Similarly, the entries of $\mathbf{b}^{(k)}_{\text{local},i}$ are given by, for $i = 1, 2, \ldots, 6$,

$$b^{(k)}_{\text{local},i} = \int_{\triangle} f(x,y) \phi_{\text{local},i} \det(F) \, \mathrm{d}A_X. \tag{10.24}$$

Entry $A^{(k)}_{\text{local},i,j}$, $i = 1, 2, \ldots, 6$, $j = 1, 2, \ldots, 6$, contributes to entry A_{k_i,k_j} of A, and entry $b^{(k)}_{\text{local},i}$, $i = 1, 2, \ldots, 6$, contributes to entry b_{k_i} of $\mathbf{b}$.

We now need to write each term in the integrands of Eqs. (10.23) and (10.24) as a function of X and Y. Remembering from Eq. (7.37) that

$$F = \begin{pmatrix} \frac{\partial x}{\partial X} & \frac{\partial x}{\partial Y} \\ \frac{\partial y}{\partial X} & \frac{\partial y}{\partial Y} \end{pmatrix} = \begin{pmatrix} x_{k_2} - x_{k_1} & x_{k_3} - x_{k_1} \\ y_{k_2} - y_{k_1} & y_{k_3} - y_{k_1} \end{pmatrix}, \tag{10.25}$$

we may easily evaluate the determinant of F. Further, Eq. (7.40) allows us to write

$$\begin{pmatrix} \frac{\partial X}{\partial x} & \frac{\partial X}{\partial y} \\ \frac{\partial Y}{\partial x} & \frac{\partial Y}{\partial y} \end{pmatrix} = F^{-1},$$

allowing us to use the chain rule to evaluate partial derivatives of the basis local functions with respect to x and y. These partial derivatives are given by, for $i = 1, 2, \ldots, 6$:

$$\frac{\partial \phi_{\text{local},i}}{\partial x} = \frac{\partial \phi_{\text{local},i}}{\partial X} \frac{\partial X}{\partial x} + \frac{\partial \phi_{\text{local},i}}{\partial Y} \frac{\partial Y}{\partial x},$$
$$\frac{\partial \phi_{\text{local},i}}{\partial y} = \frac{\partial \phi_{\text{local},i}}{\partial X} \frac{\partial X}{\partial y} + \frac{\partial \phi_{\text{local},i}}{\partial Y} \frac{\partial Y}{\partial y}.$$

Functions such as $p(x, y)$ may be written as functions of X and Y by using Eqs. (10.5) and (10.6) to first write x and y as functions of X and Y.

All terms in the integrands of Eqs. (10.23) and (10.24) may now written as a function of X and Y. The integrals may then be evaluated, either analytically, or using the quadrature rules over the canonical triangle derived in Sect. 8.7.

10.5.2 Evaluating Integrals Over Ω_N

In contrast to the integrals over element edges described in Sect. 8.6.2, the edges of elements will now contain three nodes. Suppose that edge k of the element edges that lie on $\partial\Omega_N$, denoted by ∂e_{N_k}, starts at node k_1, passes through node k_2 and ends at node k_3. Only the basis functions ϕ_{k_1}, ϕ_{k_2} and ϕ_{k_3} will be nonzero on this edge. As a consequence, the contribution to Eq. (10.22) from integration over this edge will generate nonzero contributions only to the entries b_{k_1}, b_{k_2} and b_{k_3} of the global vector $\mathbf{b}$. We store only these three nonzero local contributions from integrating over this edge in the vector $\mathbf{b}^{(k)}_{\text{local edge}}$ with entries given by, for $i = 1, 2, 3$,

$$b^{(k)}_{\text{local edge},i} = \int_{\partial e_{N_k}} \phi_{k_i}(x, y) g_N(x, y)\, \mathrm{d}s. \tag{10.26}$$

Entry $b^{(k)}_{\text{local edge},i}$, $i = 1, 2, 3$, of these local contributions will contribute to the entry b_{k_i} of the global vector $\mathbf{b}$. As the edge ∂e_{N_k} connects node k_1 to node k_3, and the edge is a straight line, points x and y that lie on this edge are parameterised by, for $0 \le X \le 1$,

$$x = (1 - X)x_{k_1} + X x_{k_3}, \tag{10.27}$$

$$y = (1 - X)y_{k_1} + X y_{k_3}. \tag{10.28}$$

Further, we may write the basis functions ϕ_{k_1}, ϕ_{k_2} and ϕ_{k_3} in terms of the variable X as, for $0 \le X \le 1$,

$$\phi_{k_1}(X) = 2\left(\frac{1}{2} - X\right)(1 - X),$$

$$\phi_{k_2}(X) = 4X(1 - X),$$

$$\phi_{k_3}(X) = 2X\left(X - \frac{1}{2}\right).$$

It is straightforward to justify these expressions for ϕ_{k_1}, ϕ_{k_2} and ϕ_{k_3}—it can be seen that they are quadratic functions and that they take the correct values at the node k_1 (where $X = 0$), the node k_2 (where $X = 0.5$) and the node k_3 (where $X = 1$). The integral given by Eq. (10.26) may then be written as, for $i = 1, 2, 3$:

$$b^{(k)}_{\text{local edge},i} = h \int_0^1 \phi_{k_i}(X) g_N(x, y)\, \mathrm{d}X, \tag{10.29}$$

where h is the length of the edge, given by

$$h = \sqrt{\left(x_{k_1} - x_{k_3}\right)^2 + \left(y_{k_1} - y_{k_3}\right)^2},$$

and x and y are given as functions of X by Eqs. (10.27) and (10.28).

If the integral given by Eq. (10.29) cannot be evaluated analytically, then it may be evaluated using quadrature techniques, as described in Sect. 3.5. On each local edge, having calculated the local contributions, we then add these contributions into the global vector $\mathbf{b}$.

10.5.3 Setting Entries Defined Explicitly

Finally, we set the entries of A and $\mathbf{b}$ given by Eqs. (10.19) and (10.20). At this stage, it is important to correct any incorrect entries that have been introduced earlier in the assembly process—see Sect. 8.6.3 for more details.

10.6 Exercises

The exercises below implement the material described in this chapter. Should the reader wishes to apply these techniques to other boundary value problems, then suitable examples are given in the exercises at the end of Chap. 8.

10.1 In Listing 8.1, we provided a computational implementation for calculating the finite element solution, using linear basis functions on a mesh of triangular elements, of the boundary value problem defined, for $-1 < x < 1, -1 < y < 1$, by

$$-\nabla \cdot \left(\mathrm{e}^x \nabla u\right) + 3x\frac{\partial u}{\partial x} + 2y\frac{\partial u}{\partial y} - 6u = -2\mathrm{e}^x\left(1 + x + 6y\right),$$

subject to Dirichlet boundary conditions

$$\begin{aligned} u(x,-1) &= x^2 - 2, && -1 < x < 1, \\ u(1,y) &= 2y^3 + 1, && -1 < y < 1, \end{aligned}$$

and natural Neumann boundary conditions

$$\begin{aligned} \mathrm{e}^x\left(\nabla u\right)\cdot\mathbf{n} &= 6\mathrm{e}^x, && -1 < x < 1, && y = 1, \\ \mathrm{e}^x\left(\nabla u\right)\cdot\mathbf{n} &= 2\mathrm{e}^x, && x = -1, && -1 < y < 1. \end{aligned}$$

This boundary value problem has solution $u(x, y) = x^2 + 2y^3$.

In this exercise, we will modify Listing 8.1 so that it calculates the finite element solution of this boundary value problem using quadratic basis functions.

(a) The function `GenerateMesh` contained in Listing 8.1 generates the coordinates of nodes, the connectivity array $lnods$, a list of nodes on the Dirichlet boundary and an array that stores the nodes that lie on each element edge of the boundary $\partial\Omega_N$. Replace the function `GenerateMesh` contained in Listing 8.1 by a new function that generates these arrays for quadratic basis functions.

(b) Modify the function `CalculateContributionOnElement` so that it computes the local contributions to the linear system that arise from integrating over element k, $A^{(k)}_{\text{local}}$ and $\mathbf{b}^{(k)}_{\text{local}}$, when quadratic basis functions are used. When modifying this function, you should remember that: (i) there are now six nodes in the local element, and so there will be six local basis functions, and $A^{(k)}_{\text{local}}$ and $\mathbf{b}^{(k)}_{\text{local}}$ will be of a different size; (ii) expressions for the local basis functions will be different to the original listing; and (iii) the partial derivatives of the basis functions will no longer be constant over the element.

(c) Modify the function `CalculateContributionOnNeumannEdge` so that it now calculates the contribution from integrating over an element edge that lies on $\partial\Omega_N$ when quadratic basis functions are used.

(d) Make any other modifications to Listing 8.1 that are needed, such as changing the size of the global matrix A and global vector $\mathbf{b}$ in lines 16–17.

Verify that your implementation is correct by comparison with the true solution.

10.2 A cubic approximation on the canonical triangular element $X \geq 0$, $Y \geq 0$, $1 - X - Y \geq 0$ may be written

$$a_1 + a_2x + a_3y + a_4x^2 + a_5xy + a_6y^2 + a_7x^3 + a_8x^2y + a_9xy^2 + a_{10}y^3,$$

and so we have ten degrees of freedom on each element. We therefore need to define ten local basis functions on the canonical triangular element. Ten nodes on this element are shown in Fig. 10.5, together with a local numbering of nodes. Suitable local basis functions are then

$$\phi_{\text{local},1}(X,Y) = \frac{9}{2}\left(\frac{1}{3} - X - Y\right)\left(\frac{2}{3} - X - Y\right)(1 - X - Y),$$

$$\phi_{\text{local},2}(X,Y) = \frac{9}{2}X\left(X - \frac{1}{3}\right)\left(X - \frac{2}{3}\right),$$

$$\phi_{\text{local},3}(X,Y) = \frac{9}{2}Y\left(Y - \frac{1}{3}\right)\left(Y - \frac{2}{3}\right),$$

$$\phi_{\text{local},4}(X,Y) = \frac{27}{2}XY\left(X - \frac{1}{3}\right),$$

$$\phi_{\text{local},5}(X,Y) = \frac{27}{2}XY\left(Y - \frac{1}{3}\right),$$

$$\phi_{\text{local},6}(X,Y) = \frac{27}{2}Y\left(Y - \frac{1}{3}\right)(1 - X - Y),$$

$$\phi_{\text{local},7}(X,Y) = \frac{27}{2}Y\left(\frac{2}{3} - X - Y\right)(1 - X - Y),$$

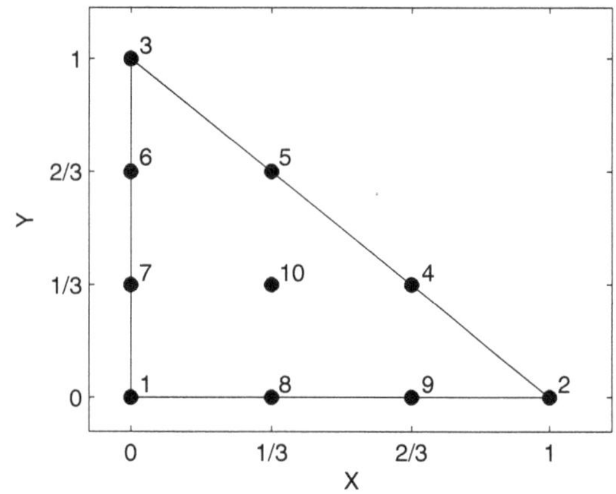

Fig. 10.5 The nodes for cubic basis functions on the canonical triangular element

$$\phi_{\text{local},8}(X, Y) = \frac{27}{2}X\left(\frac{2}{3} - X - Y\right)(1 - X - Y),$$
$$\phi_{\text{local},9}(X, Y) = \frac{27}{2}X\left(X - \frac{1}{3}\right)(1 - X - Y),$$
$$\phi_{\text{local},10}(X, Y) = 27XY(1 - X - Y).$$

(a) Generate a mesh that partitions the region $0 < x < 1, 0 < y < 1$ into triangular elements. Generate nodes so that each element has nodes that may be mapped to the nodes of the canonical element shown in Fig. 10.5.
(b) The model boundary value problem in Chap. 7 was defined by

$$-\left(\frac{\partial^2 u}{\partial x^2} + \frac{\partial^2 u}{\partial y^2}\right) = 1, \qquad 0 < x < 1, \quad 0 < y < 1,$$

subject to Dirichlet boundary conditions on the whole boundary, given by

$$\begin{aligned} u(x, 0) &= 0, & 0 < x < 1 \\ u(1, y) &= 0, & 0 < y < 1, \\ u(x, 1) &= 0, & 0 < x < 1, \\ u(0, y) &= 0, & 0 < y < 1. \end{aligned}$$

Use the mesh generated in part (a) to calculate the finite element solution of this boundary value problem that makes a cubic approximation to the solution on each element. Validate your solution by comparison with the output of Listing 7.1.

10.3 A biquadratic approximation on the canonical quadrilateral element $0 \leq X \leq 1, 0 \leq Y \leq 1$, may be written

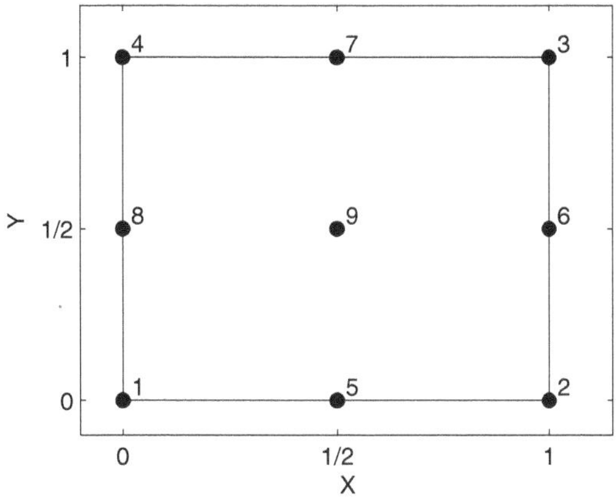

Fig. 10.6 The nodes for biquadratic basis functions on quadrilateral elements

$$a_1 + a_2x + a_3y + a_4x^2 + a_5xy + a_6y^2 + a_7x^2y + a_8xy^2 + a_9x^2y^2,$$

and so we have nine degrees of freedom on each element. We therefore need to define nine local basis functions on the canonical element $0 \leq X \leq 1, 0 \leq Y \leq 1$. Nine nodes on this element are shown in Fig. 10.6, together with a local numbering of nodes. Defining

$$\begin{aligned} Q_1(Z) &= 2\left(\frac{1}{2} - Z\right)(1 - Z), \\ Q_2(Z) &= 4Z(1 - Z), \\ Q_3(Z) &= 2Z\left(Z - \frac{1}{2}\right), \end{aligned}$$

Suitable local basis functions are then

$$\begin{aligned} \phi_{\text{local},1}(X, Y) &= Q_1(X)Q_1(Y), \\ \phi_{\text{local},2}(X, Y) &= Q_3(X)Q_1(Y), \\ \phi_{\text{local},3}(X, Y) &= Q_3(X)Q_3(Y), \\ \phi_{\text{local},4}(X, Y) &= Q_1(X)Q_3(Y), \\ \phi_{\text{local},5}(X, Y) &= Q_2(X)Q_1(Y), \\ \phi_{\text{local},6}(X, Y) &= Q_3(X)Q_2(Y), \\ \phi_{\text{local},7}(X, Y) &= Q_2(X)Q_3(Y), \\ \phi_{\text{local},8}(X, Y) &= Q_1(X)Q_2(Y), \\ \phi_{\text{local},9}(X, Y) &= Q_2(X)Q_2(Y). \end{aligned}$$

(a) Generate a mesh that partitions the region $0 < x < 1$, $0 < y < 1$ into quadrilateral elements. Generate nodes so that each element has nodes that may be mapped to the nodes of the canonical element shown in Fig. 10.6.
(b) Use the mesh generated in part (a) to calculate the finite element solution to the boundary value problem given in Exercise 10.2, where your finite element solution makes a biquadratic approximation to the solution on each element. Validate your solution by comparison with the output of Listing 7.1.

Chapter 11
Nonlinear Elliptic Partial Differential Equations

In Chap. 5, we explained how to apply the finite element method to nonlinear ordinary differential equations. We saw that calculating the finite element solution of nonlinear differential equations required us to solve a nonlinear system of algebraic equations and discussed how these algebraic equations could be solved.

In this chapter, we explain how to apply the finite element method to nonlinear partial differential equations by combining: (i) the material on calculating the finite element solution of linear partial differential equations given in Chaps. 7–10 and (ii) the material on calculating the finite element solution of nonlinear ordinary differential equations in Chap. 5. We consider equations of the form

$$-\nabla \cdot \left(p\left(x, y, u, \frac{\partial u}{\partial x}, \frac{\partial u}{\partial y}\right)\right) + r\left(x, y, u, \frac{\partial u}{\partial x}, \frac{\partial u}{\partial y}\right) = 0, \qquad (x, y) \in \Omega,$$

where Ω is some subset of the (x, y)-plane, subject to suitable boundary conditions on the boundary $\partial\Omega$. The application of the finite element method to equations of this form is illustrated using an example.

11.1 A Model Problem

We illustrate the application of the finite element method to nonlinear elliptic partial differential equations using the model differential equation, defined for the region Ω defined by $0 < x < 1$, $0 < y < 1$, by

$$-\nabla \cdot \left(\mathrm{e}^{u}\nabla u\right) + \mathrm{e}^{x^2+y^2}\left(\left(\frac{\partial u}{\partial x}\right)^2 + \left(\frac{\partial u}{\partial y}\right)^2 + 4\right) = 0, \qquad (11.1)$$

J. Whiteley, *Finite Element Methods*, Mathematical Engineering,
DOI 10.1007/978-3-319-49971-0_11

subject to Dirichlet boundary conditions

$$u(x, 0) = x^2, \qquad 0 < x < 1, \tag{11.2}$$
$$u(x, 1) = x^2 + 1, \qquad 0 < x < 1, \tag{11.3}$$
$$u(0, y) = y^2, \qquad 0 < y < 1, \tag{11.4}$$

and the Neumann boundary condition

$$e^u (\nabla u) \cdot \mathbf{n} = 2e^{y^2+1}, \qquad x = 1, \quad 0 < y < 1, \tag{11.5}$$

where $\mathbf{n}$ is the normal vector pointing out of Ω, with unit length. This boundary value problem has solution $u(x, y) = x^2 + y^2$.

11.2 The Weak Formulation

The weak formulation of the differential equation given by Eq. (11.1), subject to the boundary conditions given by Eqs. (11.2)–(11.5), is obtained by first multiplying Eq. (11.1) by a test function $v(x, y) \in H_0^1(\Omega)$ and integrating the resulting product over the region Ω:

$$\int_\Omega v \left(-\nabla \cdot \left(e^u \nabla u \right) + e^{x^2+y^2} \left(\left(\frac{\partial u}{\partial x} \right)^2 + \left(\frac{\partial u}{\partial y} \right)^2 + 4 \right) \right) \mathrm{d}A = 0.$$

After a little manipulation, followed by the application of the divergence theorem, this becomes

$$- \int_{\partial\Omega} v e^u (\nabla u) \cdot \mathbf{n} \, \mathrm{d}s + \int_\Omega e^u (\nabla u) \cdot (\nabla v) + v e^{x^2+y^2} \left(\left(\frac{\partial u}{\partial x} \right)^2 + \left(\frac{\partial u}{\partial y} \right)^2 + 4 \right) \mathrm{d}A = 0,$$

where $\partial\Omega$ is the boundary of Ω, and $\mathbf{n}$ is the normal vector of unit length pointing out of Ω. As $v(x, y) \in H_0^1(\Omega)$, then, by definition, $v(x, y) = 0$ on $\partial\Omega_D$, the part of the boundary where Dirichlet boundary conditions are applied. Using this observation, and the Neumann boundary condition given by Eq. (11.5), we may write the integral above as

$$\int_\Omega e^u (\nabla u) \cdot (\nabla v) + v e^{x^2+y^2} \left(\left(\frac{\partial u}{\partial x} \right)^2 + \left(\frac{\partial u}{\partial y} \right)^2 + 4 \right) \mathrm{d}A - \int_{\partial\Omega_N} 2e^{y^2+1} v \, \mathrm{d}s = 0,$$

where $\partial\Omega_N$ is the part of the boundary where Neumann boundary conditions are applied, namely $x = 1, 0 < y < 1$.

The weak solution is then defined by:

find $u(x, y) \in H_E^1(\Omega)$ such that

$$\int_{\Omega} \mathrm{e}^u (\nabla u) \cdot (\nabla v) + v\mathrm{e}^{x^2+y^2} \left(\left(\frac{\partial u}{\partial x} \right)^2 + \left(\frac{\partial u}{\partial y} \right)^2 + 4 \right) \mathrm{d}A - \int_{\partial\Omega_N} 2\mathrm{e}^{y^2+1} v \,\mathrm{d}s = 0, \tag{11.6}$$

for all test functions $v(x, y) \in H_0^1(\Omega)$.

11.3 The Mesh and Basis Functions

We partition Ω into a mesh of triangular elements as described in Sect. 7.3.1 and calculate a finite element solution that uses the linear basis functions described in Sect. 7.3.2. The sets S^h, S_E^h and S_0^h are then defined by Eqs. (7.22)–(7.24).

11.4 The Finite Element Solution

We write the finite element solution as the member of S^h given by

$$U(x, y) = \sum_{j=1}^{N_{\text{node}}} U_j \phi_j(x, y), \tag{11.7}$$

where N_{node} is the number of nodes in the mesh. Using the definition of the weak solution given by Eq. (11.6), we define the finite element solution by

find $U(x, y) \in S_E^h$ such that

$$\int_{\Omega} \mathrm{e}^U (\nabla U) \cdot (\nabla \phi_i) + \phi_i \mathrm{e}^{x^2+y^2} \left(\left(\frac{\partial U}{\partial x} \right)^2 + \left(\frac{\partial U}{\partial y} \right)^2 + 4 \right) \mathrm{d}A - \int_{\partial\Omega_N} 2\mathrm{e}^{y^2+1} \phi_i \,\mathrm{d}s = 0, \tag{11.8}$$

for all test functions $\phi_i(x, y) \in S_0^h$.

11.5 The Nonlinear System of Algebraic Equations

When applying the finite element method to linear differential equations, whether ordinary differential equations or partial differential equations, the definition of the finite element solution allowed us to derive a linear system of algebraic equations. In Chap. 5, we saw that, for nonlinear ordinary differential equations, the definition of the finite element solution resulted in a system of nonlinear algebraic equations that we write as

$$\mathbf{R}(\mathbf{U}) = \mathbf{0}, \tag{11.9}$$

where $\mathbf{R}$ is a vector of length N_{node}. As with linear differential equations, the individual equations that comprise Eq. (11.9) fall into two categories: (i) equations that arise from forcing $U(x, y)$ to satisfy the Dirichlet boundary conditions and lie in the set S^h_E; and (ii) equations that arise from substituting a suitable basis function $\phi_i(x, y)$ into Eq. (11.8). We now discuss these categories separately.

11.5.1 Satisfying the Dirichlet Boundary Conditions

We may combine the Dirichlet boundary conditions given by Eqs. (11.2)–(11.4) into the single boundary condition

$$u(x, y) = g_D(x, y), \qquad (x, y) \in \Omega_D,$$

where $g_D(x, y)$ is a suitable encapsulation of the right-hand sides of Eqs. (11.2)–(11.4). Suppose node i lies on the part of the boundary where Dirichlet boundary conditions are applied. Using Eq. (7.26), we may write $U(x_i, y_i) = U_i$, and so the Dirichlet boundary condition is satisfied at this node provided

$$U_i = g_D(x_i, y_i).$$

Entry i of the residual vector $\mathbf{R}$ that appears in Eq. (11.9) is therefore given by

$$R_i = U_i - g_D(x_i, y_i). \tag{11.10}$$

11.5.2 Using Suitable Test Functions

In Sect. 11.5.1, we derived an expression for entry R_i of the residual $\mathbf{R}$ that appears in Eq. (11.9) when node i lies on $\partial\Omega_D$. Suppose, instead, that node i does not lie on $\partial\Omega_D$. The basis function $\phi_i(x, y)$ is then a member of the set S^h_0 and is a suitable test function for Eq. (11.8). When node i does not lie on $\partial\Omega_D$, we then have

$$R_i = \int_{\Omega} e^U (\nabla U) \cdot (\nabla \phi_i) + \phi_i e^{x^2+y^2} \left(\left(\frac{\partial U}{\partial x} \right)^2 + \left(\frac{\partial U}{\partial y} \right)^2 + 4 \right) dA - \int_{\partial\Omega_N} 2e^{y^2+1} \phi_i \, ds. \tag{11.11}$$

In common with the examples in earlier chapters, we will find it convenient to decompose the integral over Ω above into the sum of contributions from individual elements and to decompose the integral over Ω_N into the sum of contributions from individual element edges. We therefore write

$$R_i = \sum_{k=1}^{N_{\text{ele}}} \int_{e_k} \left(e^U \left(\frac{\partial U}{\partial x} \frac{\partial \phi_i}{\partial x} + \frac{\partial U}{\partial y} \frac{\partial \phi_i}{\partial y} \right) + \phi_i e^{x^2+y^2} \left(\left(\frac{\partial U}{\partial x} \right)^2 + \left(\frac{\partial U}{\partial y} \right)^2 + 4 \right) \right) dA - \sum_{k=1}^{N_{\text{neu}}} \int_{\partial e_{N_k}} 2e^{y^2+1} \phi_i \, ds, \tag{11.12}$$

where N_{ele} is the number of elements in the mesh, e_k is the area occupied by element k, N_{neu} is the number of element edges that lie on $\partial\Omega_N$, and ∂e_{N_k} is element edge k that lies on $\partial\Omega_N$.

11.6 Assembling the Nonlinear System of Algebraic Equations

To use the finite element solution given by Eq. (11.7), we first have to solve the system of nonlinear algebraic equations, Eq. (11.9), with entries given by Eqs. (11.10) and (11.11). In Chap. 5, we saw that computational methods exist for solving systems of nonlinear algebraic equations that require only the specification of the residual vector $\mathbf{R}$. These methods are, however, inefficient for systems with a large number of unknowns. Under these conditions, more efficient computational methods should be used that require, in addition to the residual vector $\mathbf{R}$, the Jacobian matrix J, with entries given by

$$J_{i,j} = \frac{\partial R_i}{\partial U_j}. \tag{11.13}$$

We now explain how both $\mathbf{R}$ and J may be evaluated in practice. As before, this is done by: (i) evaluating integrals over Ω, (ii) evaluating integrals over $\partial\Omega_N$ and (iii) evaluating contributions that can be set explicitly. We now give some more details on each of these tasks.

11.6.1 Evaluating Integrals Over Ω

As we are using a mesh of triangular elements, and a linear approximation to the solution on each element, each element will contain three nodes. Suppose k_1, k_2, k_3 are the nodes of element k, listed in an anticlockwise direction. Only the basis functions ϕ_{k_1}, ϕ_{k_2}, ϕ_{k_3} will be nonzero on this element, and so the only nonzero contributions to Eq. (11.12) from this element arise from $i = k_1$, $i = k_2$ and $i = k_3$. We therefore calculate only these nonzero entries and store them in a local residual vector $\mathbf{R}_{\text{local}}^{(k)}$, with entries given by:

$$R_{\text{local},i}^{(k)} = \int_{e_k} \left(e^U \left(\frac{\partial U}{\partial x}\frac{\partial \phi_{k_i}}{\partial x} + \frac{\partial U}{\partial y}\frac{\partial \phi_{k_i}}{\partial y} \right) + \phi_{k_i} e^{x^2+y^2} \left(\left(\frac{\partial U}{\partial x}\right)^2 + \left(\frac{\partial U}{\partial y}\right)^2 + 4 \right) \right) \mathrm{d}A, \qquad i = 1, 2, 3. \quad (11.14)$$

Entry i, where $i = 1, 2, 3$, of $\mathbf{R}_{\text{local}}^{(k)}$, then contributes to entry k_i of the global residual $\mathbf{R}$.

For a given $U(x, y)$, the local residual vector discussed above may be calculated by mapping element k to the canonical triangle using the transformation given by Eqs. (7.12) and (7.13). The terms in the integrand may all be written as functions of X and Y on the canonical element—Eqs. (7.17)–(7.19) allow us to write

$$\phi_{k_i}(X, Y) = \phi_{\text{local},i}(X, Y), \qquad i = 1, 2, 3,$$
$$U(X, Y) = U_{k_1}\phi_{\text{local},1}(X, Y) + U_{k_2}\phi_{\text{local},2}(X, Y) + U_{k_3}\phi_{\text{local},3}(X, Y). \quad (11.15)$$

Finally, Eqs. (7.12) and (7.13) allow x and y to be written as functions of X and Y, allowing us to evaluate the terms that include x and y. We may now evaluate the integral given by Eq. (11.14), either analytically, or using quadrature as described in Sect. 8.7.

We now turn our attention to computing the entries of the Jacobian matrix defined by Eq. (11.13). Let node i be a node of the mesh that does not lie on $\partial\Omega_D$. Using Eq. (11.12), entries of the Jacobian matrix in row i may be written, for $j = 1, 2, \ldots, N_{\text{node}}$,

$$\frac{\partial R_i}{\partial U_j} = \sum_{k=1}^{N_{\text{ele}}} \int_{e_k} \frac{\partial}{\partial U_j} \left(\mathrm{e}^U \left(\frac{\partial U}{\partial x}\frac{\partial \phi_i}{\partial x} + \frac{\partial U}{\partial y}\frac{\partial \phi_i}{\partial y} \right) + \phi_i \mathrm{e}^{x^2+y^2} \left(\left(\frac{\partial U}{\partial x}\right)^2 + \left(\frac{\partial U}{\partial y}\right)^2 + 4 \right) \right) \mathrm{d}A, \quad (11.16)$$

and so entries depend only on integrals over individual elements, and there is no contribution from integrals over element edges that lie on $\partial\Omega_N$. We now consider contributions to the Jacobian matrix from the same element k, with nodes k_1, k_2, k_3, as used above when evaluating local contributions to the residual vector. We immediately see that contributions from element k to the Jacobian matrix are nonzero only for the three rows $i = k_1$, $i = k_2$ and $i = k_3$. Furthermore, we see from Eq. (11.15), that the finite element solution $U(x, y)$ on element k depends on only three values of $\mathbf{U}$: the values U_j, with $j = k_1$, $j = k_2$ and $j = k_3$. The partial derivative with respect to U_j that appears in Eq. (11.16) is therefore zero unless j takes one of the values k_1, k_2, k_3. As such, we may store all nonzero local contributions to J from element k in the 3×3 matrix $J^{(k)}_{\text{local}}$, with entries given by, for $i = 1, 2, 3$, $j = 1, 2, 3$:

$$J^{(k)}_{\text{local},i,j} = \int_{e_k} \frac{\partial}{\partial U_{k_j}} \left(\mathrm{e}^U \left(\frac{\partial U}{\partial x}\frac{\partial \phi_{k_i}}{\partial x} + \frac{\partial U}{\partial y}\frac{\partial \phi_{k_i}}{\partial y} \right) + \phi_{k_i} \mathrm{e}^{x^2+y^2} \left(\left(\frac{\partial U}{\partial x}\right)^2 + \left(\frac{\partial U}{\partial y}\right)^2 + 4 \right) \right) \mathrm{d}A.$$

This integral may be evaluated by mapping to the canonical triangle, and evaluating all quantities as functions of X and Y as was done for the local contribution to the residual. The local contributions are then added to the appropriate entry of the global Jacobian matrix; for $i = 1, 2, 3$, $j = 1, 2, 3$, the entry $J^{(k)}_{\text{local},i,j}$ of the local Jacobian matrix contributes to the entry J_{k_i,k_j} of the global Jacobian matrix.

11.6.2 Evaluating Integrals Over $\partial\Omega_N$

The contributions to Eq. (11.12) from integrals over element edges that lie on $\partial\Omega_N$ take exactly the same form as the integrals evaluated in Sect. 8.6.2. These contributions may, therefore, be evaluated in exactly the same way. We noted in the previous section that there is no contribution to the Jacobian matrix from integrating over $\partial\Omega_N$.

11.6.3 Setting Entries Defined Explicitly

The entries of **R** that take the form of Eq. (11.10) may easily be evaluated explicitly. The entries of the Jacobian matrix that arise from these rows are given by:

$$J_{i,j} = \begin{cases} 1, & i = j, \\ 0, & i \neq j, \end{cases}$$

and so these entries may also be set explicitly. Remember, however, to remove any incorrect entries that were earlier introduced into these rows when integrating over Ω and $\partial\Omega_N$. See Sect. 8.6.3 for more details.

11.7 Exercises

11.1 Use the material in this chapter to calculate the finite element solution of the differential equation given by Eq. (11.1), subject to the boundary conditions given by Eqs. (11.2)–(11.5). Verify that your finite element solution is correct by comparison with the true solution $u(x, y) = x^2 + y^2$.

11.2 Calculate the finite element solution of the differential equation, defined for $0 < x < 1, 0 < y < 1$, by:

$$-\nabla \cdot \left((1 + u^2)\nabla u\right) + (x^2 + y^2)u + 3\left(\left(\frac{\partial u}{\partial x}\right)^2 + \left(\frac{\partial u}{\partial y}\right)^2\right)u = 0,$$

subject to Dirichlet boundary conditions

$$\begin{aligned} u(x, 0) &= 1, & 0 < x < 1, \\ u(1, y) &= \mathrm{e}^y, & 0 < y < 1, \\ u(x, 1) &= \mathrm{e}^x, & 0 < x < 1, \\ u(0, y) &= 1, & 0 < y < 1. \end{aligned}$$

Use your own choice of triangular elements or quadrilateral elements and your own choice of polynomial approximation on each element. Verify that your finite element solution is correct by comparison with the true solution $u(x, y) = \mathrm{e}^{xy}$.

Chapter 12
Systems of Elliptic Equations

We saw, in Chap. 6, how the finite element method may be applied to systems of ordinary differential equations. Then, in Chaps. 7–11, we saw how to apply the finite element method to elliptic partial differential equations, using a variety of meshes and basis functions. In this chapter, we combine this material, allowing us to apply the finite element method to systems of elliptic partial differential equations.

12.1 A Model Problem

We define Ω to be the region $0 < x < 1$, $0 < y < 1$. Our model problem is the system of partial differential equations, defined on Ω by:

$$-\nabla \cdot (\nabla u_1 + \nabla u_2) + 3u_1 + 3u_2 = 2\mathrm{e}^{x+y}, \tag{12.1}$$

$$-\nabla \cdot (2\nabla u_1 + \nabla u_2) + 3u_1 = -\mathrm{e}^{x}\left(3\mathrm{e}^{y} - \mathrm{e}^{-y}\right), \tag{12.2}$$

subject to Dirichlet boundary conditions

$$u_1(x,0) = 2\mathrm{e}^{x}, \qquad u_2(x,0) = 0, \qquad 0 < x < 1, \tag{12.3}$$

$$u_1(1,y) = 2\mathrm{e}\cosh y, \qquad u_2(1,y) = 2\mathrm{e}\sinh y, \qquad 0 < y < 1, \tag{12.4}$$

$$u_1(x,1) = 2\mathrm{e}^{x}\cosh 1, \qquad u_2(x,1) = 2\mathrm{e}^{x}\sinh 1, \qquad 0 < x < 1, \tag{12.5}$$

$$u_2(0,y) = 2\sinh y, \qquad 0 < y < 1, \tag{12.6}$$

and the Neumann boundary condition

$$(\nabla u_1 + \nabla u_2) \cdot \mathbf{n} = -2\mathrm{e}^{y}, \qquad x = 0,\ 0 < y < 1, \tag{12.7}$$

J. Whiteley, *Finite Element Methods*, Mathematical Engineering,
DOI 10.1007/978-3-319-49971-0_12

where $\mathbf{n}$ is the normal vector pointing out of Ω, of unit length. This model problem has true solution

$$u_1(x, y) = 2e^x \cosh y, \tag{12.8}$$
$$u_2(x, y) = 2e^x \sinh y. \tag{12.9}$$

12.2 The Weak Formulation

We will follow a similar method to derive the weak formulation as we did in Sect. 6.2 for systems of ordinary differential equations. We multiply the first differential equation, Eq. (12.1), by a suitable test function $v_1(x, y)$, and the second equation, Eq. (12.2), by another suitable test function $v_2(x, y)$. Before we do this, we must decide which sets v_1 and v_2 belong to. This requires us to allocate the boundary conditions, given by Eqs. (12.3)–(12.7), to each of the two differential equations.

The Neumann boundary condition given by Eq. (12.7) is a natural boundary condition for the first differential equation, Eq. (12.1). If we combine this boundary condition with the Dirichlet boundary conditions on u_1 given by Eqs. (12.3)–(12.5), we then have boundary conditions on the whole of the boundary $\partial\Omega$ for Eq. (12.1). We then allocate the Dirichlet boundary conditions on u_2 given by Eqs. (12.3)–(12.6) to the second partial differential equation, Eq. (12.2). We now have boundary conditions for both equations on the whole boundary $\partial\Omega$.

The first differential equation, Eq. (12.1), satisfies Neumann boundary conditions on $x = 0$, $0 < y < 1$, and Dirichlet boundary conditions on the remainder of the boundary. The test function $v_1(x, y)$ then belongs to the set $H^1(\Omega)$, subject to the condition that $v_1(x, y) = 0$ where Dirichlet boundary conditions are applied for Eq. (12.1). Performing the usual tasks of multiplying Eq. (12.1) by $v_1(x, y)$, integrating the resulting product over Ω, applying the divergence theorem and using the Neumann boundary condition given by Eq. (12.7), we obtain:

$$\int_\Omega (\nabla u_1) \cdot (\nabla v_1) + (\nabla u_2) \cdot (\nabla v_1) + 3u_1 v_1 + 3u_2 v_1 \, \mathrm{d}A = \int_\Omega 2e^{x+y} v_1 \, \mathrm{d}A - \int_{\partial\Omega_N} 2e^y v_1 \, \mathrm{d}s, \tag{12.10}$$

where $\partial\Omega_N$ is the boundary $x = 0, 0 < y < 1$ where Neumann boundary conditions are applied.

The second differential equation, Eq. (12.2), satisfies Dirichlet boundary conditions on the whole boundary $\partial\Omega$. The test function $v_2(x, y)$ therefore belongs in the set $H^1(\Omega)$, subject to the condition that $v_2(x, y) = 0$ on $\partial\Omega$. Performing the same operations as were used to derive Eq. (12.10) yields

$$\int_{\Omega} 2\,(\nabla u_1)\cdot(\nabla v_2) + (\nabla u_2)\cdot(\nabla v_2) + 3u_1 v_2 \,\mathrm{d}A = \\ -\int_{\Omega} \mathrm{e}^x \left(3\mathrm{e}^y - \mathrm{e}^{-y}\right) v_2 \,\mathrm{d}A. \tag{12.11}$$

The weak solutions, $u_1(x, y)$ and $u_2(x, y)$, are then defined by:

find $u_1(x, y)$, $u_2(x, y)$ that satisfy the Dirichlet boundary conditions and are such that

$$\int_{\Omega} (\nabla u_1)\cdot(\nabla v_1) + (\nabla u_2)\cdot(\nabla v_1) + 3u_1 v_1 + 3u_2 v_1 \,\mathrm{d}A = \\ \int_{\Omega} 2\mathrm{e}^{x+y} v_1 \,\mathrm{d}A - \int_{\partial\Omega_N} 2\mathrm{e}^{y} v_1 \,\mathrm{d}s, \tag{12.12}$$

$$\int_{\Omega} 2\,(\nabla u_1)\cdot(\nabla v_2) + (\nabla u_2)\cdot(\nabla v_2) + 3u_1 v_2 \,\mathrm{d}A = \\ -\int_{\Omega} \mathrm{e}^x \left(3\mathrm{e}^y - \mathrm{e}^{-y}\right) v_2 \,\mathrm{d}A, \tag{12.13}$$

for all test functions $v_1(x, y)$, $v_2(x, y)$ that are zero when Dirichlet boundary conditions are applied to the appropriate equation.

12.3 The Mesh and Basis Functions

We will partition Ω into a mesh of triangular elements and use the linear basis functions described in Sect. 7.3.2. The sets S^h, S^h_E and S^h_0 are then defined by Eqs. (7.22)–(7.24).

12.4 The Finite Element Solution

We write the finite element solutions as the members of S^h given by

$$U^{(1)}(x, y) = \sum_{j=0}^{N_{\text{node}}} U_j^{(1)} \phi_j(x, y), \tag{12.14}$$

$$U^{(2)}(x, y) = \sum_{j=0}^{N_{\text{node}}} U_j^{(2)} \phi_j(x, y), \tag{12.15}$$

where N_{node} is the number of nodes in the mesh, and $\phi_j(x)$, $j = 1, 2, \ldots, N_{\text{node}}$, are the basis functions defined in Sect. 7.3.2. The finite element solutions, $U^{(1)}(x, y)$ and $U^{(2)}(x, y)$, are then defined by:

find $U^{(1)}(x, y)$, $U^{(2)}(x, y)$, given by Eqs. (12.14) and (12.15), that satisfy the Dirichlet boundary conditions and are such that

$$\int_{\Omega} \left(\nabla U^{(1)}\right) \cdot (\nabla \phi_i) + \left(\nabla U^{(2)}\right) \cdot (\nabla \phi_i) + 3U^{(1)}\phi_i + 3U^{(2)}\phi_i \; \mathrm{d}A = \int_{\Omega} 2\mathrm{e}^{x+y}\phi_i \; \mathrm{d}A - \int_{\partial\Omega_N} 2\mathrm{e}^{y}\phi_i \; \mathrm{d}s, \quad (12.16)$$

for all test functions $\phi_i(x, y)$ that are zero when Dirichlet boundary conditions are applied to Eq. (12.1), and

$$\int_{\Omega} 2\left(\nabla U^{(1)}\right) \cdot (\nabla \phi_i) + \left(\nabla U^{(2)}\right) \cdot (\nabla \phi_i) + 3U^{(1)}\phi_i \; \mathrm{d}A = -\int_{\Omega} \mathrm{e}^{x}\left(3\mathrm{e}^{y} - \mathrm{e}^{-y}\right)\phi_i \; \mathrm{d}A, \quad (12.17)$$

for all test functions $\phi_i(x, y)$ that are zero when Dirichlet boundary conditions are applied to Eq. (12.2).

12.5 The Algebraic Equations

We now derive the system of algebraic equations that arises from substituting the expressions for $U^{(1)}(x, y)$, $U^{(2)}(x, y)$, given by Eqs. (12.14) and (12.15), into the conditions required from the finite element solution that were stated in the previous section.

Suppose node i of the mesh lies on the part of the boundary where Dirichlet boundary conditions are applied for the first differential equation, Eq. (12.1). To satisfy the Dirichlet boundary conditions, we must set

$$U_i^{(1)} = g_{D_1}(x_i, y_i), \quad (12.18)$$

where $g_{D_1}(x, y)$ is the function that encapsulates the Dirichlet boundary conditions on u_1 given by Eqs. (12.3)–(12.5). If node i does not lie on the part of the boundary where Dirichlet boundary conditions are applied for the first differential equation, then $\phi_i(x, y)$ will be zero on this boundary and is a suitable basis function for substituting into Eq. (12.16). Using the expressions for $U^{(1)}(x, y)$, $U^{(2)}(x, y)$, given by Eqs. (12.14) and (12.15), we obtain, after some manipulation:

$$\sum_{j=1}^{N_{\text{node}}} \left(\int_{\Omega} (\nabla \phi_j) \cdot (\nabla \phi_i) + 3\phi_i \phi_j \, \mathrm{d}A \right) U_j^{(1)} +$$
$$\sum_{j=1}^{N_{\text{node}}} \left(\int_{\Omega} (\nabla \phi_j) \cdot (\nabla \phi_i) + 3\phi_j \phi_i \, \mathrm{d}A \right) U_j^{(2)} =$$
$$\int_{\Omega} 2\mathrm{e}^{x+y} \phi_i \, \mathrm{d}A - \int_{\partial \Omega_N} 2\mathrm{e}^{y} \phi_i \, \mathrm{d}s, \tag{12.19}$$

We may combine Eqs. (12.18)–(12.19) into the equation

$$A^{(11)}\mathbf{U}^{(1)} + A^{(12)}\mathbf{U}^{(2)} = \mathbf{b}^{(1)}, \tag{12.20}$$

where if node i of the mesh lies on the part of the boundary where Dirichlet boundary conditions are applied for the first differential equation, then

$$\begin{aligned} A_{i,j}^{(11)} &= \begin{cases} 1, & j = i, \\ 0, & j \neq i, \end{cases} \\ A_{i,j}^{(12)} &= 0, \\ b_i^{(1)} &= g_{D_1}(x_i, y_i), \end{aligned}$$

and if node i does not lie on this boundary, then

$$\begin{aligned} A_{i,j}^{(11)} &= \int_{\Omega} (\nabla \phi_j) \cdot (\nabla \phi_i) + 3\phi_i \phi_j \, \mathrm{d}A, \\ A_{i,j}^{(12)} &= \int_{\Omega} (\nabla \phi_j) \cdot (\nabla \phi_i) + 3\phi_j \phi_i \, \mathrm{d}A, \\ b_i^{(1)} &= \int_{\Omega} 2\mathrm{e}^{x+y} \phi_i \, \mathrm{d}A - \int_{\partial \Omega_N} 2\mathrm{e}^{y} \phi_i \, \mathrm{d}s. \end{aligned}$$

Applying a similar approach for the second differential equation, Eq. (12.2), we obtain, after much manipulation, the equation

$$A^{(21)}\mathbf{U}^{(1)} + A^{(22)}\mathbf{U}^{(2)} = \mathbf{b}^{(2)}, \tag{12.21}$$

where if node i of the mesh lies on the part of the boundary where Dirichlet boundary conditions are applied for the second differential equation, then

$$A_{i,j}^{(21)} = 0,$$
$$A_{i,j}^{(22)} = \begin{cases} 1, & j = i, \\ 0, & j \neq i, \end{cases}$$
$$b_i^{(2)} = g_{D_2}(x_i, y_i),$$

where $g_{D_2}(x, y)$ is the function that encapsulates the Dirichlet boundary conditions on u_2 given by Eqs. (12.3)–(12.6). If node i does not lie on this boundary, then

$$A_{i,j}^{(21)} = \int_{\Omega} 2\left(\nabla\phi_j\right) \cdot (\nabla\phi_i) + 3\phi_i\phi_j \, \mathrm{d}A,$$
$$A_{i,j}^{(22)} = \int_{\Omega} \left(\nabla\phi_j\right) \cdot (\nabla\phi_i) \, \mathrm{d}A,$$
$$b_i^{(2)} = -\int_{\Omega} \mathrm{e}^x\left(3\mathrm{e}^y - \mathrm{e}^{-y}\right)\phi_i \, \mathrm{d}A.$$

We then compute the vectors $\mathbf{U}^{(1)}$ and $\mathbf{U}^{(2)}$ by combining Eqs. (12.20) and (12.21) into the single linear system

$$A\mathbf{U} = \mathbf{b},$$

where

$$A = \begin{pmatrix} A^{(11)} & A^{(12)} \\ A^{(21)} & A^{(22)} \end{pmatrix}, \qquad \mathbf{U} = \begin{pmatrix} \mathbf{U}^{(1)} \\ \mathbf{U}^{(2)} \end{pmatrix}, \qquad \mathbf{b} = \begin{pmatrix} \mathbf{b}^{(1)} \\ \mathbf{b}^{(2)} \end{pmatrix}.$$

This linear system may be assembled by calculating the entries as described in Chaps. 7 and 8.

12.6 Exercises

12.1 Using the material presented in this chapter, calculate the finite element solution of the system of equations given by Eqs. (12.1)–(12.7), using a mesh of triangular elements and making a linear approximation to the solution on each element.

12.2 Extend the computational implementation that you wrote in Exercise 12.1 to use, for example, a quadrilateral mesh and/or higher order basis functions.

Chapter 13
Parabolic Partial Differential Equations

We now describe how to apply the finite element to parabolic partial differential equations. This is done by approximating the parabolic partial differential equation by either a sequence of ordinary differential equations or a sequence of elliptic partial differential equations. We may then solve these ordinary differential equations or elliptic partial differential equations using the techniques developed earlier in this book.

13.1 A Linear Parabolic Partial Differential Equation

Our first example is the parabolic partial differential equation, defined for $0 < x < 1$, $0 < y < 1$, $0 < t < 3$, by

$$\frac{\partial u}{\partial t} = \nabla \cdot (\nabla u) + f(x, y, t), \tag{13.1}$$

where

$$f(x, y, t) = \mathrm{e}^{-t}\left(2x(1-x) + 2y(1-y) - xy(1-x)(1-y)\right),$$

subject to the Dirichlet boundary conditions given by, for $0 < t < 3$:

$$u(x, 0, t) = 0, \qquad 0 < x < 1, \tag{13.2}$$

$$u(1, y, t) = 0, \qquad 0 < y < 1, \tag{13.3}$$

$$u(x, 1, t) = 0, \qquad 0 < x < 1, \tag{13.4}$$

$$u(0, y, t) = 0, \qquad 0 < y < 1, \tag{13.5}$$

J. Whiteley, *Finite Element Methods*, Mathematical Engineering,
DOI 10.1007/978-3-319-49971-0_13

and the initial conditions

$$u(x, y, 0) = xy(1 - x)(1 - y), \qquad 0 < x < 1, \qquad 0 < y < 1. \tag{13.6}$$

This differential equation has solution $u(x, y, t) = xy(1 - x)(1 - y)\mathrm{e}^{-t}$.

We begin by discretising Eq. (13.1) in time. Let $0 = t_0 < t_1 < t_2 < \cdots < t_{M-1} < t_M = 3$ be the values of t where we approximate $u(x, y, t)$. We then define, for $m = 0, 1, 2, \ldots, M$, the function $u^{(m)}(x, y)$ to be the approximation to the function $u(x, y, t)$ evaluated at $t = t_m$, i.e.,

$$u^{(m)}(x, y) \approx u(x, y, t_m). \tag{13.7}$$

The function $u^{(m)}(x, y)$ is often called "the solution on timestep m". We place the superscript m in parentheses in Eq. (13.7) to avoid confusion with u being raised to the power of m. Clearly, from the initial conditions given by Eq. (13.6), we have

$$u^{(0)}(x, y) = xy(1 - x)(1 - y), \qquad 0 < x < 1, \qquad 0 < y < 1. \tag{13.8}$$

The boundary conditions given by Eqs. (13.2)–(13.5) imply that for $m = 1, 2, \ldots, M$,

$$u^{(m)}(x, 0) = 0, \qquad 0 < x < 1, \tag{13.9}$$

$$u^{(m)}(1, y) = 0, \qquad 0 < y < 1, \tag{13.10}$$

$$u^{(m)}(x, 1) = 0, \qquad 0 < x < 1, \tag{13.11}$$

$$u^{(m)}(0, y) = 0, \qquad 0 < y < 1. \tag{13.12}$$

The partial derivative of u with respect to time on the interval $t_{m-1} < t \leq t_m$, $m = 1, 2, \ldots, M$, may be approximated using

$$\frac{\partial u}{\partial t} \approx \frac{u^{(m)}(x, y) - u^{(m-1)}(x, y)}{t_m - t_{m-1}}, \qquad t_{m-1} < t \leq t_m.$$

In our finite element approximation, on the interval $t_{m-1} < t \leq t_m$, $m = 1, 2, \ldots, M$, we evaluate all terms on the right-hand side of Eq. (13.1) at time t_m. This is known as an *implicit approximation* to the differential equation. An implicit approximation ensures that the finite element solutions computed have good numerical stability properties; i.e., this approximation ensures that the inevitable small rounding errors that occur in any computational implementation do not grow exponentially, eventually swamping the solution. On this interval, we may then write

$$\frac{\partial^2 u}{\partial x^2} = \frac{\partial^2 u^{(m)}}{\partial x^2}, \qquad \frac{\partial^2 u}{\partial y^2} = \frac{\partial^2 u^{(m)}}{\partial y^2}.$$

An implicit discretisation in time of Eq. (13.1) on the interval $t_{m-1} < t \leq t_m$, $m = 1, 2, \ldots, M$, then satisfies, for $0 < x < 1, 0 < y < 1$,

$$\frac{u^{(m)} - u^{(m-1)}}{t_m - t_{m-1}} = \frac{\partial^2 u^{(m)}}{\partial x^2} + \frac{\partial^2 u^{(m)}}{\partial y^2} + f(x, y, t_m),$$

which, after a little rearranging, may be written as

$$-\left(\frac{\partial^2 u^{(m)}}{\partial x^2} + \frac{\partial^2 u^{(m)}}{\partial y^2}\right) + \frac{u^{(m)}}{t_m - t_{m-1}} = f(x, y, t_m) + \frac{u^{(m-1)}}{t_m - t_{m-1}}. \quad (13.13)$$

Provided we know $u^{(m-1)}(x, y)$, Eqs. (13.9)–(13.13) are a linear elliptic partial differential equation and boundary conditions for $u^{(m)}(x, y)$, $m = 1, 2, \ldots, M$. A finite element solution to this boundary value problem may easily be computed using the finite element techniques given in Chaps. 7–10. As $u^{(0)}(x, y)$ is given by Eq. (13.8), we may use Eqs. (13.9)–(13.13) to compute the finite element solution $U^{(1)}(x, y)$ that is an approximation to $u^{(1)}(x, y)$. This finite element approximation $U^{(1)}(x)$ may be used, along with Eqs. (13.9)–(13.13), to compute the finite element solution $U^{(2)}(x, y)$ that is an approximation to $u^{(2)}(x, y)$. We then proceed in this manner, calculating a finite element solution $U^{(m)}(x, y)$ that is an approximation to $u^{(m)}(x, y)$, until we reach the final time where $m = M$.

We leave the computational implementation of the calculation of this finite element solution as an exercise for the reader. There is, however, one very important point to be made on the finite element solution of the sequence of elliptic partial differential equations described above, which has a significant effect on the efficiency of the computational implementation. When solving each linear elliptic partial differential equation given by Eq. (13.13), where $m = 1, 2, \ldots, M$, the finite element solution on timestep m will require the solution of the linear system

$$A^{(m)}\mathbf{U}^{(m)} = \mathbf{b}^{(m)}. \quad (13.14)$$

Suppose the points $t_0, t_1, \ldots, t_M$ are equally spaced, so that $t_m - t_{m-1}$ takes a constant value for $m = 1, 2, \ldots, M$. It will be seen in Exercise 13.1 that $A^{(m)}$ is identical on each timestep, and so this matrix only need be computed once, at the start of the computational implementation.

13.2 A Nonlinear Parabolic Equation

An example nonlinear parabolic equation is Fisher's equation:

$$\frac{\partial u}{\partial t} = \frac{\partial^2 u}{\partial x^2} + u(1 - u), \qquad -1 < x < 10, \quad 0 < t < 10, \quad (13.15)$$

subject to boundary conditions

$$\frac{\partial u}{\partial x} = 0, \qquad x = -1, 10, \tag{13.16}$$

and initial conditions

$$u(x, 0) = u_0(x), \qquad -1 < x < 10, \tag{13.17}$$

where

$$u_0(x) = \begin{cases} 1, & -1 < x \leq 0, \\ 0, & 0 < x < 10. \end{cases}$$

As with the linear differential equation discussed in Sect. 13.1, we let $0 = t_0 < t_1 < t_2 < \cdots < t_{M-1} < t_M = 10$ be the values of t where we approximate $u(x, t)$ and then define, for $m = 0, 1, 2, \ldots, M$,

$$u^{(m)}(x) \approx u(x, t_m).$$

The initial conditions given by Eq. (13.17) may then be written as

$$u^{(0)}(x) = u_0(x), \tag{13.18}$$

and the boundary conditions given by Eq. (13.16) imply that

$$\frac{\mathrm{d}u^{(m)}}{\mathrm{d}x} = 0, \qquad x = -1, 10. \tag{13.19}$$

An implicit discretisation in time of Eq. (13.15) on the interval $t_{m-1} < t \leq t_m$, $m = 1, 2, \ldots, M$ then satisfies

$$\frac{u^{(m)} - u^{(m-1)}}{t_m - t_{m-1}} = \frac{\mathrm{d}^2 u^{(m)}}{\mathrm{d}x^2} + u^{(m)}(1 - u^{(m)}), \qquad -1 < x < 10,$$

which, after a little rearranging, may be written as

$$-\frac{\mathrm{d}^2 u^{(m)}}{\mathrm{d}x^2} - u^{(m)}(1 - u^{(m)}) + \frac{u^{(m)}(x) - u^{(m-1)}}{t_m - t_{m-1}} = 0, \quad -1 < x < 10. \tag{13.20}$$

Provided we know $u^{(m-1)}(x)$, the differential equation given by Eq. (13.20) is a nonlinear differential equation for $u^{(m)}$, subject to the boundary conditions given by Eq. (13.19). We may calculate the finite element approximation $U^{(m)}$ using the techniques discussed in Chap. 5 for nonlinear ordinary differential equations. We may therefore proceed through all timesteps m as in Sect. 13.1, calculating finite element solutions $U^{(m)}$ that approximate the functions $u^{(m)}(x)$, until we reach the final time where $m = M$.

13.3 A Semi-implicit Discretisation in Time

We have used an implicit approximation to the time derivative in all examples pre sented so far in this chapter. We explained earlier that this ensured that the finite element solutions calculated had good numerical stability properties. There are some equations where the so-called semi-implicit approximation is more computationally efficient and also possesses good stability properties. Suppose we are solving the reaction-diffusion system given, for $a < x < b$, $0 < t < T$, by

$$\frac{\partial u}{\partial t} = D\nabla \cdot (\nabla u) + f(x, y, t, u), \tag{13.21}$$

where D is constant, subject to suitable initial and boundary conditions. Calculating the finite element solution at times $0 = t_0 < t_1 < t_2 < \cdots < t_{M-1} < t_M = T$, and writing $u^{(m)}(x) \approx u(x, t_m)$ for $m = 0, 1, 2, \ldots, M$, the implicit approximation of Eq. (13.21) is a nonlinear differential equation for $u^{(m)}$, $m = 1, 2, \ldots, M$ given by

$$-D\nabla \cdot \left(\nabla u^{(m)}\right) + \frac{u^{(m)} - u^{(m-1)}}{t_m - t_{m-1}} - f(x, y, t_m, u^{(m)}) = 0.$$

The semi-implicit approximation evaluates the term given by $f(x, y, t, u)$ at time t_{m-1} rather than time t_m. The semi-implicit approximation may then be written as

$$- D\nabla \cdot \left(\nabla u^{(m)}\right) + \frac{u^{(m)}}{t_m - t_{m-1}} = \frac{u^{(m-1)}}{t_m - t_{m-1}} + f(x, y, t_{m-1}, u^{(m-1)}). \tag{13.22}$$

The semi-implicit approximation given by Eq. (13.22) has two attractive features. First, the differential equation for $u^{(m)}$, given by Eq. (13.22), is a linear equation. The finite element solution can therefore be calculated by solving a linear system of algebraic equations rather than a nonlinear system. Secondly, provided the timestep $t_m - t_{m-1}$ is constant for $m = 1, 2, ..., M$, the matrix that appears in the linear system is the same for each timestep, as discussed in Sect. 13.1. Thus, the semi-implicit approximation has the computational advantage that this matrix need only be calculated once. Although the stability properties of the semi-implicit method are not as good as the implicit method, this very rarely causes a problem.

13.4 Exercises

13.1 Calculate the finite element solution of the parabolic partial differential equation given by Eqs. (13.1)–(13.6). Validate your finite element solution by comparison with the true solution $u(x, y, t) = xy(1 - x)(1 - y)e^{-t}$.

On each timestep, you need to solve a linear system. Use a constant timestep, so that $t_m - t_{m-1}$ is constant for $m = 1, 2, ..., M$. Verify that for this model problem, you need only calculate the matrix that appears in this linear system once, before looping over timesteps.

13.2 A parabolic partial differential equation in one spatial dimension is defined by, for $0 < x < \pi, 0 < t < 10$:

$$\frac{\partial u}{\partial t} = \frac{\partial^2 u}{\partial x^2} + \sqrt{2} \sin\left(x + t + \frac{1}{4}\pi\right),$$

with boundary conditions given by

$$\begin{aligned} u &= \sin t, & x &= 0, & 0 &< t < 10, \\ \frac{\partial u}{\partial x} &= \cos(\pi + t), & x &= \pi, & 0 &< t < 10, \end{aligned}$$

and initial conditions given by

$$u(x, 0) = \sin x, \qquad 0 < x < \pi.$$

This differential equation has solution $u(x, t) = \sin(x + t)$.

(a) Using a uniform mesh for both the spatial region $0 < x < \pi$ and the time interval $0 < t < 10$ calculate the finite element approximation to the solution at time $t = 10$. Use a linear approximation to the solution on each element. Note that you only need to calculate the global finite element matrix once, as described in Sect. 13.1.
(b) Repeat part (a) using a higher order polynomial approximation on each element.

13.3 Fisher's equation is given by Eqs. (13.15)–(13.17).

(a) Calculate the finite element solution of this nonlinear partial differential equation using the implicit method described in Sect. 13.2.
(b) Calculate the finite element solution of this nonlinear partial differential equation using the semi-implicit method described in Sect. 13.3.

13.4 A system of parabolic partial differential equations, in one spatial dimension, that describes pattern formation (see Murray [5], for more details) is given by

$$\frac{\partial u_1}{\partial t} = \frac{\partial^2 u_1}{\partial x^2} + \gamma \left(a - u_1 - \frac{\rho u_1 u_2}{1 + u_1 + K u_1^2} \right),$$

$$\frac{\partial u_2}{\partial t} = d\frac{\partial^2 u_2}{\partial x^2} + \gamma \left(\alpha(b - u_2) - \frac{\rho u_1 u_2}{1 + u_1 + K u_1^2} \right),$$

for $0 < x < 10$, $0 < t < 50$, with parameter values $\alpha = 1.5$, $\gamma = 9$, $K = 0.1$, $\rho = 18.5$, $a = 92$, $b = 64$, $d = 10$, initial conditions $u_1(x, 0) = 10 + \sin x$, $u_2(x, 0) = 9 + \sin x$, and boundary conditions

$$\frac{\partial u_1}{\partial x} = \frac{\partial u_2}{\partial x} = 0, \quad x = 0, 10, \qquad 0 < t < 50.$$

Calculate the finite element solution of this system of equations. This system may be solved using the semi-implicit approximation described in Sect. 13.3.

13.5 The system of parabolic partial differential equations given in Exercise 13.4 may be generalised to two dimensions by writing them as, for $0 < t < 50$,

$$\frac{\partial u_1}{\partial t} = \nabla \cdot (\nabla u_1) + \gamma \left(a - u_1 - \frac{\rho u_1 u_2}{1 + u_1 + K u_1^2} \right),$$

$$\frac{\partial u_2}{\partial t} = d\nabla \cdot (\nabla u_2) + \gamma \left(\alpha(b - u_2) - \frac{\rho u_1 u_2}{1 + u_1 + K u_1^2} \right),$$

using the same parameter values as in Exercise 13.4, initial conditions $u_1(x, y, 0) = 10 + \sin x + \sin y$, $u_2(x, y, 0) = 9 + \sin x + \sin y$, and boundary conditions

$$\mathbf{n} \cdot \nabla u_1 = \mathbf{n} \cdot \nabla u_2 = 0,$$

on all boundaries for $0 < t < 50$.

(a) Solve the partial differential equations above on the square $0 < x < 10$, $0 < y < 1$, for times $0 < t < 50$. This system may be solved using the semi-implicit approximation described in Sect. 13.3.
(b) Repeat part (a), with the parameter γ now taking the value $\gamma = 0.9$.

Appendix A
Methods for Solving Linear and Nonlinear Systems of Algebraic Equations

All applications of the finite element method in this book have required the solution of a system of algebraic equations. These algebraic equations were linear when the differential equation was linear, and nonlinear when the differential equation was nonlinear. Despite the central role played by the solution of these systems of equations, we have said very little about how to solve them.

The numerical solution of both linear and nonlinear systems of equations is underpinned by very elegant mathematics and is a fascinating field in its own right. There are many excellent expositions of this subject, for example Trefethen and Bau [11], Golub and van Loan [3], Greenbaum [4], van der Vorst [12] and the recent review article by Wathen [13], to name a few. A more specialised treatment from a finite element standpoint may be found in Elman, Silvester and Wathen [1]. It is not possible to do justice to this field in a short appendix, and we do not attempt to do so. Instead, we present an overview of established, practical techniques that may be used to solve the systems of algebraic equations arising from the finite element method. Although we provide an outline of the algorithms that may be used, we do not recommend that the user develops software to implement these algorithms themselves. Indeed, we would discourage the reader from doing so, as many commercial and open-source software packages (for example, MATLAB, Octave, PETSc) are readily available and provide reliable, computationally efficient and robust implementations. The purpose of outlining the algorithms here is to allow the reader to understand both the operations performed by each algorithm and—most importantly—the consequences of these operations in terms of computational requirements. It is intended that this will give the reader guidance in choosing the most appropriate technique for their computations. In the discussion that follows, we assume some familiarity with basic linear algebra and continuous mathematics, but aim to make the material as accessible as possible.

Throughout this book, we have used the convention that the finite element solution at node i in a finite element mesh is given by U_i. Using this convention, systems of algebraic equations have been written as the linear system $A\mathbf{U} = \mathbf{b}$, or the nonlinear system $\mathbf{R}(\mathbf{U}) = \mathbf{0}$. In a slight abuse of this convention, in this appendix, we write these systems of equations as $A\mathbf{u} = \mathbf{b}$ and $\mathbf{R}(\mathbf{u}) = \mathbf{0}$, respectively, i.e. we write the

J. Whiteley, *Finite Element Methods*, Mathematical Engineering,
DOI 10.1007/978-3-319-49971-0

vector of unknowns as **u** rather than **U**. This is mainly to avoid confusion with upper triangular matrices, which will be used in this chapter and are commonly denoted by U. Furthermore, writing the vector of unknowns as **u** fits in with a common convention from linear algebra, namely that vectors are represented by lower-case letters.

A.1 Linear Systems

Linear systems arising from the finite element method have two key properties: (i) they are nonsingular, and (ii) they are sparse, i.e. they have very few nonzero entries in each row of the matrix. Techniques for solving these linear systems fall into two categories, direct methods and iterative methods, that we now discuss.

A.1.1 Direct Methods

We begin by describing a method for solving linear systems that most readers will be familiar with: *Gaussian elimination* with partial pivoting, sometimes known as row pivoting. This technique, although not requiring sophisticated mathematics, is a practical solver for many linear systems. As explained above, we do not provide full details of the computational implementation of this technique, but merely enough detail to allow the reader to understand both the computational operations that are required and the practical consequences of these operations. First, we require some definitions.

A matrix is said to be *lower triangular* if all entries above the diagonal take the value zero. Clearly, the matrix L is lower triangular if, and only if, $L_{i,j} = 0$ when $i < j$. Similarly, a matrix U is *upper triangular* if, and only if, all entries below the diagonal take the value zero, that is $U_{i,j} = 0$ if $i > j$. Finally, a *permutation matrix* is a square matrix with the following two properties: (i) each row and each column of the matrix contain precisely one entry that takes the value 1, and (ii) all other entries of the matrix take the value 0. Permutation matrices take their name because they permute rows of a matrix, as we shall now explain. Suppose P is a permutation matrix of size $N \times N$ where, in row i, the entry that takes the value 1 appears in column j. Let A be another matrix of size $N \times N$ and $\mathbf{b}$ be a row vector with N entries. The entries of row j of A will appear in row i of PA, and the entry in row j of $\mathbf{b}$ will appear in row i of $P\mathbf{b}$. Permutation matrices are used to implement the pivoting steps of Gaussian elimination.

Suppose we solve the nonsingular linear system

$$A\mathbf{u} = \mathbf{b}, \tag{A.1}$$

using Gaussian elimination with partial pivoting, where A is a $N \times N$ matrix. One common practical implementation of this technique yields a permutation matrix P, a lower triangular matrix L with all diagonal entries taking the value 1 and an upper triangular matrix U with all diagonal entries nonzero, such that $PA = LU$, where all matrices are of size $N \times N$. This is often known as the "LU decomposition of A". Noting that Eq. (A.1) may be written as

$$PA\mathbf{u} = \hat{\mathbf{b}},$$

where

$$\hat{\mathbf{b}} = P\mathbf{b},$$

we see that the LU decomposition of A allows us to write the given linear system as

$$LU\mathbf{u} = \hat{\mathbf{b}}. \tag{A.2}$$

Care should be taken not to confuse the upper triangular matrix U with the vector of unknows $\mathbf{u}$. Setting $\mathbf{v} = U\mathbf{u}$ allows us to decompose the linear system as the coupled system

$$L\mathbf{v} = \hat{\mathbf{b}}, \qquad U\mathbf{u} = \mathbf{v}. \tag{A.3}$$

Although it may appear that decomposing the linear system into the two systems above doubles the work required, this is not the case. The structure of triangular matrices allows us to explicitly solve the system very simply. Remembering that L is a lower triangular matrix with all diagonal entries taking the value 1 and that U is an upper triangular matrix, we may write Eq. (A.3) as:

$$\begin{pmatrix} 1 & 0 & 0 & \dots & 0 \\ L_{2,1} & 1 & 0 & \dots & 0 \\ L_{3,1} & L_{3,2} & 1 & \dots & 0 \\ \vdots & \vdots & \vdots & \ddots & \vdots \\ L_{N,1} & L_{N,2} & L_{N,3} & \dots & 1 \end{pmatrix} \begin{pmatrix} v_1 \\ v_2 \\ v_3 \\ \vdots \\ v_N \end{pmatrix} = \begin{pmatrix} \hat{b}_1 \\ \hat{b}_2 \\ \hat{b}_3 \\ \vdots \\ \hat{b}_N \end{pmatrix}, \tag{A.4}$$

$$\begin{pmatrix} U_{1,1} & U_{1,2} & \dots & U_{1,N-2} & U_{1,N-1} & U_{1,N} \\ 0 & U_{2,2} & \dots & U_{2,N-2} & U_{2,N-1} & U_{2,N} \\ \vdots & \vdots & \ddots & \vdots & \vdots & \vdots \\ 0 & 0 & \dots & U_{N-2,N-2} & U_{N-2,N-1} & U_{N-2,N} \\ 0 & 0 & \dots & 0 & U_{N-1,N-1} & U_{N-1,N} \\ 0 & 0 & \dots & 0 & 0 & U_{N,N} \end{pmatrix} \begin{pmatrix} u_1 \\ u_2 \\ \vdots \\ u_{N-2} \\ u_{N-1} \\ u_N \end{pmatrix} = \begin{pmatrix} v_1 \\ v_2 \\ \vdots \\ v_{N-2} \\ v_{N-1} \\ v_N \end{pmatrix}. \tag{A.5}$$

We may sweep through the first of these linear systems, Eq. (A.4), solving for each value of v_i in turn:

$$
\begin{aligned}
v_1 &= \hat{b}_1, \\
v_2 &= \hat{b}_2 - L_{2,1}v_1, \\
v_3 &= \hat{b}_3 - L_{3,1}v_1 - L_{3,2}v_2, \\
\vdots &\quad \vdots \\
v_N &= \hat{b}_N - L_{N,1}v_1 - L_{N,2}v_2 - L_{N,3}v_3 - \ldots - L_{N,N-1}v_{N-1}.
\end{aligned}
$$

Similarly, as U is an upper triangular matrix, we may sweep through the second linear system, Eq. (A.5), by starting from the last equation:

$$
\begin{aligned}
u_N &= \frac{1}{U_{N,N}} v_N, \\
u_{N-1} &= \frac{1}{U_{N-1,N-1}} \left(v_{N-1} - U_{N-1,N}u_N \right), \\
u_{N-2} &= \frac{1}{U_{N-2,N-2}} \left(v_{N-2} - U_{N-2,N}u_N - U_{N-2,N-1}u_{N-1} \right), \\
\vdots &\quad \vdots \\
u_1 &= \frac{1}{U_{1,1}} \left(v_1 - U_{1,N}u_N - U_{1,N-1}u_{N-1} - \ldots - U_{1,2}u_2 \right).
\end{aligned}
$$

Note that division by the diagonal entries of U in the equations above is well defined, as we have explained that these entries are nonzero when using Gaussian elimination. This completes the solution of the original linear system, Eq. (A.1).

We now evaluate *LU* decomposition as a practical method for solving the linear systems that arise from the finite element method. As the size of the system increases, the computational work required is dominated by computing the matrices P, L, U, and the solution of Eq. (A.3) requires an insignificant computational effort in comparison. This is a very useful observation when many linear systems defined by the same matrix A are to be solved, as is the case for some of the parabolic partial differential equations seen in Chap. 13. For these systems, we need only evaluate P, L, U once: these matrices may then be reused. A further feature of *LU* decomposition to bear in mind is that although A is a sparse matrix, the computed factors L and U will not, in general, be sparse. Should N, the number of rows in the linear system, be large, then, although storing the sparse matrix A is feasible, storing the dense matrices L and U may not be possible. *LU* decomposition would then fail as a practical method for solving Eq. (A.1).

In summary, *LU* decomposition is suitable for problems with a relatively small number of unknowns. What “a small number of unknowns” actually means in this context will depend on the computer being used—it is likely that “small” may mean hundreds of thousands of modern laptop or desktop computers.

Readers may have wondered why we have not yet explained what a direct solver is. This is because it is easier to do so with the aid of an example, such as Gaussian elimination with partial pivoting. This algorithm uses prescribed steps, with a computational cost, both in terms of memory requirement and operations to be performed, that is known in advance. The algorithm then returns a solution that we may, neglecting computational rounding errors, assume to be the exact solution. These are the distinguishing features of any direct solver.

A.1.2 Iterative Methods

In the previous section, we loosely defined a direct method for solving a linear system to be a method that used prescribed steps, with computational complexity known a priori, to calculate the exact solution to a problem. An alternative family of methods are *iterative methods*, where an initial guess is made to the solution, and this guess improved iteratively until it is sufficiently good. By "sufficiently good", we usually mean that $\|A\mathbf{u} - \mathbf{b}\| < \varepsilon$ for some prescribed ε, where $\|\mathbf{v}\|$ represents the norm of a vector $\mathbf{v}$. Iterative methods are often used to solve the sparse linear systems that arise from the finite element method when the number of unknowns is too large for a direct method, such as that described in Sect. A.1.1, to be applied.

In the absence of specialist knowledge, some basic advice for using iterative methods to solve the sparse linear system $A\mathbf{u} = \mathbf{b}$ may be expressed very simply. If the matrix A is symmetric and positive definite—that is, $A = A^\top$ and $\mathbf{v}^\top A\mathbf{v} > 0$ for all nonzero vectors $\mathbf{v}$—then the conjugate gradient methodshould be used. If not the Generalised Minimal RESiduals (GMRES) technique should be used. We present the algorithm for both of these techniques to allow the reader to understand the computations underpin these techniques and their consequences in terms of computational memory and operations required. This is particularly important for the use of GMRES, as this technique requires some input from the user that can have a significant effect on the efficiency and computational memory requirements. Writing down the algorithms also allows us to outline how convergence may be accelerated by preconditioning techniques in Sect. A.1.3.

We also include a summary of two other iterative techniques: Gauss–Seidel iteration and multigrid. Gauss–Seidel iteration is very rarely used as an iterative technique. It is, however, often used as part of a multigrid technique. Multigrid techniques are a family of iterative techniques for solving linear systems. It is not possible, in a short appendix, to fully explain how these techniques may be applied as an effective iterative solver. Instead, we give an overview of the principles that underpin this technique. Although we would not suggest that a novice uses a basic multigrid technique as an iterative solver, we will claim, when discussing preconditioning in Sect. A.1.3, that a basic multigrid technique can be an effective preconditioner.

A.1.2.1 The Conjugate Gradient Method

The conjugate gradient method may be used for solving linear systems $A\mathbf{u} = \mathbf{b}$ when the matrix A is symmetric and positive definite. Care should be taken to avoid destroying the symmetry of A that arises in finite element applications by modifying A to satisfy Dirichlet boundary conditions. The conjugate gradient method is set out in Algorithm 1. Each step simply requires the evaluation of an algebraic expression. Vectors $\mathbf{q}_k$, $\mathbf{u}_k$ and $\mathbf{r}_k$ are created on step k of the algorithm. If these vectors were stored until the algorithm terminates, this could require substantial computational storage should the number of iterations becomes large. Note, however, that these vectors are only required for one further step of the iteration and may then be deleted. The conjugate gradient method is therefore very easy to implement and requires only minimal computational storage. It is also very easy to use—it requires only an initial guess $\mathbf{u}_0$ and tolerance ε. It can also be shown that this method has exceptional convergence properties.

Algorithm 1 The conjugate gradient algorithm

1: Enter with initial guess $\mathbf{u}_0$, tolerance $\varepsilon > 0$
2: $\mathbf{r}_0 = A\mathbf{u}_0 - \mathbf{b}$
3: $\mathbf{q}_0 = \mathbf{r}_0$
4: $k = 1$
5: **while** $\|\mathbf{r}_{k-1}\| > \varepsilon$ **do**
6: $\quad \alpha_k = -\frac{\mathbf{r}_{k-1}^\top \mathbf{r}_{k-1}}{\mathbf{q}_{k-1}^\top A\mathbf{q}_{k-1}}$
7: $\quad \mathbf{u}_k = \mathbf{u}_{k-1} + \alpha_k \mathbf{q}_{k-1}$
8: $\quad \mathbf{r}_k = \mathbf{r}_{k-1} + \alpha_k A\mathbf{q}_{k-1}$
9: $\quad \beta_k = \frac{\mathbf{r}_k^\top \mathbf{r}_k}{\mathbf{r}_{k-1}^\top \mathbf{r}_{k-1}}$
10: $\quad \mathbf{q}_k = \mathbf{r}_k + \beta_k \mathbf{q}_{k-1}$
11: $\quad k = k + 1$
12: **end while**
13: $\mathbf{u} = \mathbf{u}_{k-1}$

A.1.2.2 The Generalised Minimal RESiduals (GMRES) Method

We saw in Sect. A.1.2.1 that the conjugate gradient algorithm required only the evaluation of algebraic expressions that were straightforward to execute and did not require substantial computational memory to store the results of these computations. This makes the conjugate gradient technique a very attractive iterative method for solving linear systems. However, this method may only be used for matrices that are both symmetric definite and positive definite. A popular iterative technique for matrices that do not have these properties is the GMRES method.

The steps required for the GMRES method are given in Algorithm 2. In contrast to the conjugate gradient algorithm, there are two additional complexities. First, we see that a dense matrix H is generated as part of this algorithm. On iteration j, this matrix has $j + 1$ rows and j columns. If the number of iterations required is sufficiently small, this will not cause a problem. However, if a large number of iterations are required, this matrix will require substantial storage space in the computer's memory, which may not be possible. Furthermore, in line 15 of the algorithm, we see that we have to solve a least squares problem during each iteration of the algorithm which, in contrast to the operations required for the conjugate gradient method, is not a simple algebraic operation.

Algorithm 2 The GMRES algorithm

1: Enter with initial guess $\mathbf{u}_0$, tolerance $\varepsilon > 0$
2: $\mathbf{r}_0 = A\mathbf{u}_0 - \mathbf{b}$
3: $\beta = \|\mathbf{r}_0\|$
4: $\mathbf{f}_1 = \mathbf{r}_0/\beta$
5: $j = 1$
6: **while** $\|\mathbf{r}_{j-1}\| > \varepsilon$ **do**
7: $\quad\mathbf{w} = -A\mathbf{f}_j$
8: $\quad$**for** $i = 1, 2, \dots, j$ **do**
9: $\qquad H_{i,j} = \mathbf{w} \cdot \mathbf{f}_i$
10: $\qquad \mathbf{w} = \mathbf{w} - H_{i,j}\mathbf{f}_i$
11: $\quad$**end for**
12: $\quad H_{j+1,j} = \|\mathbf{w}\|$
13: $\quad \mathbf{f}_{j+1} = \frac{1}{H_{j+1,j}}\mathbf{w}$
14: $\quad \mathbf{e}_1 = \left(1\ 0\ 0\ \dots\ 0\right)^\top$
15: $\quad$Solve the least squares problem to find $\mathbf{y}_j$ such that $\mathbf{y}_j = \min_{\mathbf{y}} \|\beta\mathbf{e}_1 - H\mathbf{y}\|^2$
16: $\quad F = \left(\mathbf{f}_1\ \mathbf{f}_1\ \dots\ \mathbf{f}_j\right)$
17: $\quad \mathbf{u}_j = \mathbf{u}_0 + F\mathbf{y}_j$
18: $\quad \mathbf{r}_j = A\mathbf{u}_j - \mathbf{b}$
19: $\quad j = j + 1$
20: **end while**
21: $\mathbf{u} = \mathbf{u}_{j-1}$

In practice, generation of a large, dense matrix H may be avoided by limiting the maximum number of times the `while` loop is executed to some fixed number M that is specified by the user. If convergence has not been reached after M iterations of this `while` loop, the algorithm exits with final iterate $\mathbf{u}$. The GMRES algorithm is then executed again, using this computed vector $\mathbf{u}$ as the initial guess. This is repeated until convergence is obtained. Although this avoids the generation of a large, dense matrix, there is no guarantee that this algorithm will terminate, and so a suitable value of M—often known as the *restart value*—is estimated by trial and error. If GMRES performs poorly for a given choice of M, then increasing M may improve performance. This is not always true, however: see [2], for some counter-intuitive observations on the choice of restart value.

We noted above that the GMRES algorithm requires the solution of a least squares problem during each iteration of the algorithm. Least squares problems such as these are typically solved using a QR factorisation of the matrix H, that is, computing a factorisation $H = QR$ where Q is an orthogonal matrix[1] and R is an upper triangular matrix. This factorisation may be calculated using Gram–Schmidt orthogonalisation,

---

[1] An orthogonal matrix is a square matrix that satisfies $QQ^\top = I$ where I is the identity matrix.

a technique that may be implemented efficiently but is unstable to rounding errors. When H is of a small size, these rounding errors usually have little effect, and this may be the most appropriate method for computing the QR factorisation of H. The numerical instabilities may be eliminated, however, by using the modified Gram–Schmidt orthogonalisation. This technique does, however, have the drawback of being less efficient to implement. For more details, see Trefethen and Bau [11].

In summary, the GMRES method is a very useful iterative technique as it may be applied to a general linear system. The user should be aware, however, that they must specify both the restart value M and the method used for solving the least squares problem that is embedded within the algorithm. The ability to specify both of these features is included in all commonly used implementations of the GMRES method and so the user, after some experience, will be able to tune their application of the GMRES method to a specific linear system.

A.1.2.3 Gauss–Seidel Iteration

A further iterative technique is Gauss–Seidel iteration. As explained earlier, this technique is rarely used as an iterative solver but often forms part of a multigrid method that we describe later. Suppose we are solving the nonsingular linear system $A\mathbf{u} = \mathbf{b}$, where all diagonal entries of A are nonzero. Write $A = D + L + U$, where D is a diagonal matrix, L is a lower triangular matrix with all diagonal entries taking the value zero and U is an upper triangular matrix with all diagonal entries taking the value zero. Using these matrices, we may write the linear system to be solved as

$$(D + L)\mathbf{u} + U\mathbf{u} = \mathbf{b}.$$

This observation motivates the Gauss–Seidel algorithm given in Algorithm 3.

Algorithm 3 The Gauss–Seidel algorithm

1: Enter with initial guess $\mathbf{u}_0$, tolerance $\varepsilon > 0$
2: $\mathbf{r}_0 = A\mathbf{u}_0 - \mathbf{b}$
3: $k = 1$
4: **while** $\|\mathbf{r}_{k-1}\| > \varepsilon$ **do**
5: Solve $(D + L)\mathbf{u}_k = \mathbf{b} - U\mathbf{u}_{k-1}$
6: $\mathbf{r}_k = A\mathbf{u}_k - \mathbf{b}$
7: $k = k + 1$
8: **end while**
9: $\mathbf{u} = \mathbf{u}_{k-1}$

Noting that $D + L$ is a lower triangular matrix, we see that the linear system that is solved in line 5 of the Gauss–Seidel algorithm may be solved very easily using the method described for solving the lower triangular system in Sect. A.1.1. The computational memory requirements are also very low—the vectors $\mathbf{x}_k$ and $\mathbf{r}_k$ that are generated on each iteration only require preservation for at most one more iteration and can then be deleted. The drawback is that there is no guarantee that this algorithm will converge, and if it does converge, then convergence is usually slower than that for GMRES or the conjugate gradient method.

A.1.2.4 Multigrid Techniques

The final iterative techniques that we discuss are the family of techniques known as multigrid techniques. These techniques can be used as iterative solvers for linear systems: this was the initial purpose for developing them. We should emphasise that the discussion that follows focuses only on a rudimentary implementation of multigrid techniques and does not provide sufficient discussion to allow the implementation of multigrid as an efficient iterative solver. The purpose of this discussion will become clear in Sect. A.1.3, when we explain that a basic implementation of multigrid may be used to accelerate convergence of the GMRES algorithm presented in Sect. A.1.2.2.

As the name suggests, multigrid techniques use several computational meshes or grids. Suppose a finite element solution is required on a computational mesh, $\mathcal{M}_1$, with many nodes, and that the finite element solution on this mesh satisfies $A_1\mathbf{u}_1 = \mathbf{b}_1$, where both A_1 and $\mathbf{b}_1$ are evaluated from the mesh $\mathcal{M}_1$. Suppose further that $\mathcal{M}_1, \mathcal{M}_2, \ldots, \mathcal{M}_J$ are a sequence of successively coarser meshes and that, were we to calculate the finite element solution on the mesh $\mathcal{M}_j, j = 1, 2, \ldots, J$, we would obtain the linear system $A_j\mathbf{u}_j = \mathbf{b}_j$. We assume that, for $j = 1, 2, \ldots, J-1$, we may restrict vectors on mesh $\mathcal{M}_j$ to the coarser mesh $\mathcal{M}_{j+1}$ using

$$\mathbf{u}_{j+1} = R_j\mathbf{u}_j,$$

where $R_j, j = 1, 2, \ldots, J-1$, is a matrix known as the *restriction operator*. Similarly, we assume the existence of a matrix $P_j, j = 1, 2, \ldots, J-1$, known as the *prolongation operator*, that allows us to transfer a vector on the mesh $\mathcal{M}_{j+1}$ onto the finer mesh $\mathcal{M}_j$ using

$$\mathbf{u}_j = P_j\mathbf{u}_{j+1},$$

Our final assumption is that the coarsest mesh, $\mathcal{M}_J$, contains sufficiently few nodes that the resulting linear system can easily be solved using a direct method.

The ideas behind the basic multigrid techniques are simple. Let $\mathbf{u}_j$ be the solution of $A_j\mathbf{u}_j = \mathbf{b}_j$ for $j = 1, 2, \ldots, J$, and let $\mathbf{v}_j$ be an approximation to $\mathbf{u}_j$. We define

$$\mathbf{r}_j = A_j\mathbf{v}_j - \mathbf{b}_j, \qquad \mathbf{e}_j = \mathbf{v}_j - \mathbf{u}_j, \qquad j = 1, 2, \ldots, J.$$

Remembering that $A_j\mathbf{u}_j = \mathbf{b}_j$, we may deduce that

$$A_j\mathbf{e}_j = \mathbf{r}_j. \tag{A.6}$$

Given an initial guess on the finest mesh $\mathscr{M}_1$, we first perform a small number of Gauss–Seidel iterations on the equation $A_1\mathbf{u}_1 = \mathbf{b}_1$ using this initial guess, before computing the residual $\mathbf{r}_1$ on this fine mesh. We then restrict this residual onto the next mesh $\mathscr{M}_2$ using $\mathbf{r}_2 = R_1\mathbf{r}_1$, before performing a small number of Gauss–Seidel iterations on Eq. (A.6) to update the residual $\mathbf{r}_2$. This is repeated on the successively coarser meshes until we restrict the residual onto the coarsest mesh, $\mathscr{M}_J$, where we solve Eq. (A.6) exactly. We then proceed back through the meshes, prolongating the solution onto successively finer meshes and performing a small number of Gauss–Seidel iterations until we reach the finest mesh, $\mathscr{M}_1$, where we update our initial guess. We complete the multigrid cycle by performing a small number of Gauss–Seidel iterations of the equation $A_1\mathbf{u}_1 = \mathbf{b}_1$.

The multigrid method described above is presented in Algorithm 4. This multigrid cycle is repeated until the residual is sufficiently small. Note that the operations required are all algebraic operations that can be evaluated efficiently and require minimal computational storage when the algorithm is being performed.

Algorithm 4 One cycle of a basic multigrid algorithm

Enter with initial guess $\mathbf{u}_1$
M Gauss–Seidel iterations of $A_1\boldsymbol{\delta}_1 = \mathbf{b}_1$, with initial guess $\boldsymbol{\delta}_1 = \mathbf{u}_1$
$\mathbf{r}_1 = A_1\boldsymbol{\delta}_1 - \mathbf{b}_1$
for j = 2,3,…,J-1 **do**
    $\mathbf{b}_j = R_{j-1}\mathbf{r}_{j-1}$
    M Gauss–Seidel iterations of $A_j\boldsymbol{\delta}_j = \mathbf{b}_j$, with initial guess $\boldsymbol{\delta}_j = 0$
    $\mathbf{r}_j = A_j\boldsymbol{\delta}_j - \mathbf{b}_j$
end for
$\mathbf{b}_J = R_{J-1}\mathbf{r}_{J-1}$
Solve exactly $A_J\mathbf{d}_J = \mathbf{b}_J$
for j = J−1,J−2,…,2 **do**
    $\boldsymbol{\delta}_j = \boldsymbol{\delta}_j + P_j\mathbf{d}_{j+1}$
    M Gauss–Seidel iterations of $A_j\mathbf{d}_j = \mathbf{b}_j$, with initial guess $\mathbf{d}_j = \boldsymbol{\delta}_j$
end for
$\mathbf{u}_1 = \mathbf{u}_1 + P_1\mathbf{d}_2$
M Gauss–Seidel iterations of $A_1\boldsymbol{\delta}_1 = \mathbf{b}_1$, with initial guess $\boldsymbol{\delta}_1 = \mathbf{u}_1$
$\mathbf{u} = \boldsymbol{\delta}_1$

There is clearly much flexibility when using this technique—using either a different number of grids or a different number of Gauss–Seidel iterations are obvious candidates for improving convergence properties. Many other options also exist for accelerating convergence, of which we briefly mention only a few. Gauss–Seidel iterations were used in Algorithm 4 to smooth the error $\mathbf{e}_j$ on mesh $\mathscr{M}_j$. Other smoothers exist and may enhance the convergence properties. We also used the same number of Gauss–Seidel iterations on all meshes and on both the restricted and prolongated solutions—we may allow these to vary if we wish. Furthermore, in Algorithm 4, we iterated through meshes $\mathscr{M}_1$, $\mathscr{M}_2$ to the finest mesh $\mathscr{M}_J$ and then back along the reverse path to the finest mesh $\mathscr{M}_1$. A more flexible technique uses cycles that are not as uniform as this.

A.1.3 Preconditioning

It is now a good time to take stock of the material presented so far in this appendix. We discussed direct methods for solving linear systems in Sect. A.1.1 and proposed using these methods when the number of unknowns was sufficiently small that the memory requirements were not excessive. If the number of unknowns is too large for a direct method to be used, we proposed, in Sect. A.1.2, that an iterative method would be more appropriate. Specifically, we propose using the conjugate gradient method if the matrix is symmetric and positive definite, and the GMRES method otherwise. We have not discussed the number of iterations taken for either of these iterative techniques to converge as this requires far more detail than can be given in a book on the finite element method. Nevertheless, we do think it is appropriate to give some practical tips on a technique known as *preconditioning* for reducing both the number of iterations required and—importantly—the computation time.

When computing finite element solutions, we often need to solve sparse linear systems $A\mathbf{u} = \mathbf{b}$ using an iterative technique. If M is a nonsingular matrix, then this linear system is equivalent to

$$M^{-1}A\mathbf{u} = M^{-1}\mathbf{b}. \tag{A.7}$$

This equation is known as the *preconditioned system*, and the matrix M is known as a *preconditioner*. Suppose we choose the preconditioner so that $M = A$. This choice of preconditioner would allow us to write Eq. (A.7) as $\mathbf{u} = M^{-1}\mathbf{b}$, and we would then have solved the linear system. This choice of preconditioner is clearly not appropriate—if we were able to evaluate A^{-1}, then we may as well use a direct method to solve the original system. This observation motivates us to specify two conditions that a preconditioning matrix must satisfy. The first condition is that the product $M^{-1}A$ should be—speaking very loosely—some approximation to the identity matrix. We deliberately leave this as a very imprecise statement as, in this context, a matrix with tightly clustered eigenvalues may be considered a suitable approximation. The second condition is that the action of M^{-1} on a vector is computationally cheap to compute. This second condition requires explanation as it has significant practical implications. Note that we do not need an explicit expression for the matrix M^{-1}. Indeed, for most preconditioners that are used in practice, an expression for M^{-1} is not available. Suppose we want to compute $M^{-1}\mathbf{v}$ for some given vector $\mathbf{v}$. Let us write $\mathbf{w} = M^{-1}\mathbf{v}$. Computing $\mathbf{w}$ is then equivalent to solving the linear system

$$M\mathbf{w} = \mathbf{v}.$$

If M has been factorised as $M = LU$, where L is a lower triangular matrix, and U is an upper triangular matrix, we may easily solve this linear system using the method described in Sect. A.1.1. Evaluation of $\mathbf{w} = M^{-1}\mathbf{v}$, i.e. the action of M^{-1} on the vector $\mathbf{v}$, is therefore possible without M^{-1} being explicitly defined.

Suppose we are solving a linear system $A\mathbf{u} = \mathbf{b}$ using either the conjugate gradient method presented in Algorithm 1 or the GMRES method presented in Algorithm 2. In both algorithms, we see that the only operations involving the matrix A are matrix–vector multiplications of the form $A\mathbf{v}$. Suppose, instead, we solve the preconditioned system given by Eq. (A.7), using a preconditioner M that satisfies the conditions above. As the second of these conditions is that the action of M^{-1} on any vector is cheap to compute, we may easily compute the right-hand side of this linear system, i.e. $M^{-1}\mathbf{b}$. When using either the conjugate gradient or GMRES algorithm for the preconditioned system, we then have to evaluate matrix–vector multiplications of the form $M^{-1}A\mathbf{v}$ for any vector $\mathbf{v}$. These may also be evaluated cheaply: writing $\mathbf{w} = A\mathbf{v}$, we see that

$$M^{-1}A\mathbf{v} = M^{-1}\mathbf{w},$$

which, using our assumption on the preconditioning matrix, is cheap to compute. A preconditioner that satisfies the conditions outlined above will hopefully require fewer iterations of an iterative method, while not significantly increasing the computational cost of each iteration.

The preconditioned system given by Eq. (A.7) is known as a *left preconditioned* system. An alternative approach is the *right preconditioned system* given by

$$AM^{-1}\mathbf{v} = \mathbf{b}, \tag{A.8}$$

$$\text{where } \mathbf{u} = M^{-1}\mathbf{v}. \tag{A.9}$$

The solution of the right preconditioned system may be computed by first solving Eq. (A.8) for $\mathbf{v}$, followed by solving Eq. (A.9) for $\mathbf{u}$. The right preconditioned system has similar properties to the left preconditioned system in that it will hopefully decrease the number of iterations required without significantly adding to the computational cost of each algorithm.

Care should be taken when preconditioning the conjugate gradient method. This iterative technique is underpinned by the assumption that the matrix A is symmetric and positive definite, and these properties of A must not be destroyed by preconditioning. This may be achieved by writing the preconditioner as $M = CC^\top$ for some matrix C and then writing the system $A\mathbf{u} = \mathbf{b}$ as:

$$C^{-1}A\left(C^{-1}\right)^\top \mathbf{v} = \mathbf{b},$$

$$\text{where } \mathbf{u} = \left(C^\top\right)^{-1}\mathbf{v}.$$

We have now described what a preconditioner is and what properties it must satisfy. We have not yet proposed any practical preconditioners. We now give two examples.

A.1.3.1 ILU Preconditioning

Incomplete LU preconditioning—often known as ILU preconditioning—is a simple preconditioner that is straightforward to implement, yet can be very effective. This preconditioner generates two matrices $\hat{L}$, $\hat{U}$ that are approximations to the full LU decomposition of A discussed in Sect. A.1.1. The matrix $\hat{L}$ is a lower triangular matrix, with nonzero entries only in the same locations as the nonzero entries in the lower triangular portion of the matrix A. Similarly, the matrix $\hat{U}$ is an upper triangular matrix, with nonzero entries only in the same locations as the nonzero entries in the upper triangular portion of the matrix A. The preconditioner M is then given by $M = \hat{L}\hat{U}$. The action of M^{-1} on a given vector $\mathbf{v}$ is cheap to compute: writing $\mathbf{w} = M^{-1}\mathbf{v}$, we see that we may compute $\mathbf{w}$ by solving the linear system

$$\hat{L}\hat{U}\mathbf{w} = \mathbf{v},$$

and we saw in Eqs. (A.2) and (A.3) that systems of this form can be solved very easily.

ILU preconditioning is available in all major open-source and commercial software packages for the iterative solution of linear systems and is a very good default choice of preconditioner. See [12] for more details on the implementation of this preconditioner.

A.1.3.2 Multigrid as a Preconditioner

In Sect. A.1.2.4, we presented a basic multigrid method as an iterative solver. This iterative solver required repeated application of Algorithm 4 until the prescribed tolerance for convergence was met. We now note that this algorithm may actually form the basis of a preconditioner. For a general vector $\mathbf{v}$, we may define $M^{-1}\mathbf{v}$ to be a small number of iterations of a multigrid algorithm to the vector $\mathbf{v}$. This clearly satisfies both conditions for a preconditioner set out in Sect. A.1.3. As many iterations of the multigrid algorithm would solve the linear system, we may approximate the inverse of A by a small number of iterations of this algorithm. Furthermore, we noted in Sect. A.1.2.4 that an application of a low number of cycles of the multigrid algorithm—that is, the action of M^{-1} on $\mathbf{v}$—is computationally cheap.

Multigrid has been shown to be an effective preconditioner—see [6] for more details. It is, however, more difficult to implement than ILU preconditioning. Although it is available in some software packages, implementation of this preconditioner requires significant programming input from the user such as, for example, specification of the prolongation and restriction operators between meshes. In contrast, ILU preconditioning may be utilised in most software packages by a very small number of lines of code that are generic for any application.

A.2 Nonlinear Systems

A general nonlinear system of N algebraic equations may be written in vector form as

$$\mathbf{R}(\mathbf{u}) = \mathbf{0}, \tag{A.10}$$

where $\mathbf{u}$ is a vector of N unknowns. The most popular iterative technique for solving systems of equations such as Eq. (A.10) is known as Newton's method, which we now describe.

Suppose that $\mathbf{u}^*$ is a solution of this system, so that $\mathbf{R}(\mathbf{u}^*) = \mathbf{0}$. Note that we say "a solution" rather than "the solution"—there is no guarantee that this solution is unique or, indeed, that a solution $\mathbf{u}^*$ exists. Approximating component i of $\mathbf{R}$ as a Taylor series about a general point $\mathbf{u}$, up to and including all linear terms in $\mathbf{u}^* - \mathbf{u}$, gives

$$R_i(\mathbf{u}^*) \approx R_i(\mathbf{u}) + \sum_{j=1}^{N} J_{ij} \left(u_j^* - u_j \right), \qquad i = 1, 2, \ldots, N,$$

where J denotes the Jacobian matrix, with entries given by

$$J_{ij} = \left. \frac{\partial R_i}{\partial u_j} \right|_{\mathbf{u}},$$

where the solid vertical line with subscript $\mathbf{u}$ indicates that entries of J are evaluated at $\mathbf{u}$. We may write this in vector form as

$$\mathbf{R}(\mathbf{u}^*) \approx \mathbf{R}(\mathbf{u}) + J \left(\mathbf{u}^* - \mathbf{u} \right).$$

Noting that $\mathbf{u}^*$ is a solution of Eq. (A.10), we have $\mathbf{R}(\mathbf{u}^*) = \mathbf{0}$, and assuming the matrix J is invertible, we may approximate the solution of Eq. (A.10) by

$$\mathbf{u}^* \approx \mathbf{u} - J^{-1}\mathbf{R}(\mathbf{u}).$$

This approximation to $\mathbf{u}^*$ clearly has the potential to be very crude—apart from assuming that the Jacobian matrix is invertible, we have assumed that $\mathbf{u}$ is a sufficiently good guess to the solution $\mathbf{u}^*$ that the Taylor series of $\mathbf{R}$ about $\mathbf{u}$, up to and including only linear terms, is sufficiently accurate to evaluate $\mathbf{R}(\mathbf{u}^*)$.

Newton's method is usually implemented using a modification to the approximation above. Given an initial guess to the solution $\mathbf{u}_0$, we then generate successive iterates $\mathbf{u}_1, \mathbf{u}_2, \mathbf{u}_3, \ldots$ using Algorithm 5.

It is crucial to understand that Newton's method should not be considered a bulletproof technique for solving systems of nonlinear algebraic equations. Even if the system of equations has a unique solution, there is no guarantee that Newton's method

Algorithm 5 Newton's method for solving nonlinear systems of algebraic equations

1: Enter with initial guess $\mathbf{u}_0$, tolerance $\varepsilon > 0$
2: $\mathbf{r}_0 = \mathbf{R}(\mathbf{u}_0)$
3: $k = 1$
4: **while** $\|\mathbf{r}_{k-1}\| > \varepsilon$ **do**
5: Solve $J\boldsymbol{\delta} = -\mathbf{R}(\mathbf{u}_{k-1})$ where the entries of J are given by

$$J_{ij} = \left.\frac{\partial R_i}{\partial u_j}\right|_{\mathbf{u}_{k-1}},$$

6: Set $\mathbf{u}_k = \mathbf{u}_{k-1} + s\boldsymbol{\delta}$, where $0 < s \leq 1$ is chosen to minimise $\|\mathbf{R}(\mathbf{u}_k)\|$.
7: $\mathbf{r}_k = \mathbf{R}(\mathbf{u}_k)$
8: $k = k + 1$
9: **end while**
10: $\mathbf{u} = \mathbf{u}_{k-1}$

will converge to this solution for a given initial guess. This may be illustrated using a very simple example of a nonlinear equation in one scalar variable u. The function $R(u) = u\mathrm{e}^{-u}$ has the unique solution $u = 0$ to the equation $R(u) = 0$. It is, however, fairly easy to show that Newton's method will not converge to this solution for any initial guess satisfying $u_0 \geq 1$. Furthermore, should a nonlinear equation possesses multiple roots, there is no guarantee which of these roots Newton's method will converge to (if it does indeed converge). In particular, we should not expect Newton's method to converge to the root closest to the initial guess. An excellent illustration of this point may be found in Chap. 4 of Süli and Mayers, [10], using the example of solving the equation $\mathrm{e}^z - z - 2 = 0$ for the complex number $z = x + iy$. A system of two nonlinear equations for x and y may be derived by decomposing this equation into real and imaginary parts. There are an infinite number of roots of this system of equations, and the root that Newton's method converges to for a given initial guess possesses some very unexpected properties. See Süli and Mayers, [10], for more details.

The discussion in the previous paragraph should be viewed as a brief summary of the negative aspects of Newton's method. This should be balanced by the positive aspects. Newton's method is simple to implement, applicable to general systems of equations, and is usually considered the "go-to" method for solving general nonlinear systems of algebraic equations. We emphasise that widely available open-source and commercial implementations of Newton's method are available. Many of these are reliable, robust and efficient, and so the reader need not implement these algorithms themselves. Some of these implementations require only specification of the function $\mathbf{R}(\mathbf{u})$, while some require specification of both $\mathbf{R}(\mathbf{u})$ and the Jacobian matrix J. The entries of J are partial derivatives of the entries of $\mathbf{R}(\mathbf{u})$, and the reader may prefer to evaluate these derivatives numerically. If this approach is taken, then for efficiency, the contributions to the numerical derivatives should be computed on each element as described in Sect. 5.1.7.

Appendix B
Vector Calculus

In this appendix, we give a very brief summary of the vector calculus that is required when applying the finite element method to partial differential equations in two spatial dimensions.

B.1 The Gradient Operator

The gradient operator, ∇, is defined to be the operator

$$\nabla = \begin{pmatrix} \frac{\partial}{\partial x} \\ \frac{\partial}{\partial y} \end{pmatrix}.$$

If a function $u(x, y)$ possesses first partial derivatives with respect to x and y, the *gradient of* $u(x, y)$ is defined by

$$\nabla u = \begin{pmatrix} \frac{\partial u}{\partial x} \\ \frac{\partial u}{\partial y} \end{pmatrix}.$$

Similarly, if a vector-valued function $\mathbf{F}(x, y)$, given by

$$\mathbf{F}(x, y) = \begin{pmatrix} F_1(x, y) \\ F_2(x, y) \end{pmatrix},$$

is such that $F_1(x, y)$ and $F_2(x, y)$ both possess first partial derivatives with respect to x and y, the *divergence* of $\mathbf{F}$ is defined by

J. Whiteley, *Finite Element Methods*, Mathematical Engineering,
DOI 10.1007/978-3-319-49971-0

$$\nabla \cdot \mathbf{F} = \begin{pmatrix} \frac{\partial}{\partial x} \\ \frac{\partial}{\partial y} \end{pmatrix} \cdot \begin{pmatrix} F_1(x, y) \\ F_2(x, y) \end{pmatrix}$$
$$= \frac{\partial F_1}{\partial x} + \frac{\partial F_2}{\partial y}.$$

As a consequence,

$$\nabla \cdot (\nabla u) = \begin{pmatrix} \frac{\partial}{\partial x} \\ \frac{\partial}{\partial y} \end{pmatrix} \cdot \begin{pmatrix} \frac{\partial u}{\partial x} \\ \frac{\partial u}{\partial y} \end{pmatrix}$$
$$= \frac{\partial^2 u}{\partial x^2} + \frac{\partial^2 u}{\partial y^2}. \tag{B.1}$$

The expression above is sometimes written as $\nabla^2 u$.

B.2 A Useful Vector Identity

Suppose the function $u(x, y)$ possesses all first and second partial derivatives, and the functions $v(x, y)$ and $p(x, y)$ possess all first partial derivatives with respect to both x and y. An identity used when deriving the weak solution of partial differential equations is

$$\nabla \cdot (vp\nabla u) = v\nabla \cdot (p\nabla u) + p\,(\nabla v) \cdot (\nabla u)\,. \tag{B.2}$$

B.3 The Divergence Theorem

Let Ω be a closed, bounded region in two dimensions, with piecewise smooth boundary $\partial\Omega$. Let $\mathbf{F}(x, y)$ be continuous, with continuous first partial derivatives on the region Ω. We then have

$$\int_{\Omega} \nabla \cdot \mathbf{F}\, \mathrm{d}A = \int_{\partial\Omega} \mathbf{F} \cdot \mathbf{n}\, \mathrm{d}s, \tag{B.3}$$

where $\mathbf{n}$ is the normal vector pointing out of Ω, with unit length.

Further Reading

Some additional texts that have been cited in earlier chapters have been listed below.

References

1. Elman, H., Silvester, D., Wathen, A.: Finite Elements and Fast Iterative Solvers. Oxford University Press, Oxford (2005)
2. Embree, M.: The tortoise and the hare restart GMRES. SIAM Rev. **45**, 259–266 (2003)
3. Golub, G.H., Van Loan, C.F.: Matrix Computations, 4th edn. The Johns Hopkins University Press, Baltimore (2012)
4. Greenbaum, A.: Iterative Methods for Solving Linear Systems. Society for Industrial and Applied Mathematics, Philadelphia (1997)
5. Murray, J.D.: Mathematical Biology. Springer, Berlin (1993)
6. Oosterlee, C.W., Washio, T.: An evaluation of parallel multigrid as a solver and a preconditioner for singularly perturbed problems. SIAM J. Sci. Comput. **19**, 87–110 (1998)
7. Persson, P.-O., Strang, G.: A simple mesh generator in matlab. SIAM Rev. **46**, 329–345 (2004)
8. Shewchuk, J.R.: Triangle: engineering a 2D quality mesh generator and delaunay triangulator. Lect. Notes Comput. Sci. **1148**, 203–222 (1996)
9. Shewchuk, J.R.: Delaunay refinement algorithms for triangular mesh generation. Comput. Geom. Theory Appl. **22**, 21–74 (2002)
10. Süli, E., Mayers, D.F.: An Introduction to Numerical Analysis. Cambridge University Press, Cambridge (2006)
11. Trefethen, L.N., Bau, D.: Numerical Linear Algebra. Society for Industrial and Applied Mathematics, Philadelphia (1997)
12. Van der Vorst, H.A.: Iterative Krylov Methods for Large Linear Systems. Cambridge University Press, Cambridge (2003)
13. Wathen, A.J.: Preconditioning. Acta Numer. **24**, 329–376 (2015)

J. Whiteley, *Finite Element Methods*, Mathematical Engineering,
DOI 10.1007/978-3-319-49971-0

Index

J. Whiteley, *Finite Element Methods*, Mathematical Engineering,
DOI 10.1007/978-3-319-49971-0

Lightning Source UK Ltd.
Milton Keynes UK
UKHW01n0602241018
331102UK00009B/332/P

9 783319 499703